Chipless RFID

Reza Rezaiesarlak • Majid Manteghi

Chipless RFID

Design Procedure and Detection Techniques

Springer

Reza Rezaiesarlak
Virginia Tech
Blacksburg, VA, USA

Majid Manteghi
Virginia Tech
Blacksburg, VA, USA

ISBN 978-3-319-10168-2 ISBN 978-3-319-10169-9 (eBook)
DOI 10.1007/978-3-319-10169-9
Springer Cham Heidelberg New York Dordrecht London

Library of Congress Control Number: 2014948461

Printed on acid-free paper

Springer is part of Springer Science+Business Media (www.springer.com)

Preface

With the advent of Radio Frequency Identification (RFID) technology as a contactless identification technique, more attention has been drawn to the application of RFID in a variety of commercial areas such as access management, tracking of goods and animals, toll collection, contactless payment, and so on. This has driven a huge motivation within different research and industrial groups to propose a variety of techniques to realize contactless identification. Chipless RFID is one of the current proposed candidates which can compete with the traditional barcode in terms of cost and ease of fabrication. Recently, many efforts have been made to improve the design of printable and compact chipless RFID tags and the identification techniques. In spite of the many conferences and seminars held on chipless RFID tags and many publications introduced by researchers in literature, there is a need for a reference book which covers both the tag design and detection techniques applicable in practical chipless RFID systems. The goal of this book is to provide the theoretical basis and practical techniques to design and analyze chipless RFID systems. The material presented in this book will help researchers and engineers to understand the design methodology and detection techniques based on rigorous mathematical formulation.

In contrast to conventional RFID tags, the chipless RFID tags do not carry any electronic circuitry in their structure to handle a communications protocol. This makes the chipless RFID tag a fully electromagnetic scatterer and encoder. The advantage of using this technology is the simplicity of the tag manufacturing process. Using conductive ink, a conventional inkjet printer can print the tag pattern directly onto a package or item. On the other hand, the detection technique could be especially challenging in an environment containing noise, clutter, and multiple multi-bit tags. Part of this book has been devoted to the transient scattering analysis which is beneficial both for the design methodology and the detection procedure. Employing transient scattering analysis, an efficient and effective anti-collision algorithm is developed for tag detection in the case with multiple multi-bit tags in the reader range.

Although this book has been prepared as a reference on chipless RFID systems, it can be useful in other electromagnetic-based detection scenarios such as ground penetrating radar (GPR), target identification, resonant-based detection of breast cancer, human identification, biometrics, and chipless sensors.

We would like to thank the students and professors working at the Virginia Tech Antenna Group (VTAG) for their valuable support and helpful comments during the preparation of the book. Special thanks go to Professors William Davis, Gary Brown, Warren Stutzman, Ioannis Besieris, and Dr. Taeyoung Yang for their patience during discussions on theoretical parts and measurement setups. We greatly appreciate Professor Jeffrey Reed for his helpful support through the years we worked on chipless RFID systems at VTAG. We appreciate Institute for Critical Technology and Applied Science (ICTAS) at Virginia Tech for its support. We also would like to thank Nicolas Ingvoldstad for reviewing and editing the material in the book.

Blacksburg, VA, USA Majid Manteghi
Reza Rezaiesarlak

Contents

Chapter 1
Radio Frequency Identification Systems

1.1 Introduction

Promising to replace all optical barcodes and other types of automatic contactless identification techniques, the Ultra High Frequency Radio Frequency Identification (UHF RFID) systems stimulated a large-scale movement both in industry and academia in 2002. Automatic contactless identification systems are not new technologies; barcodes and various types of magnetic coupling RFIDs have been on the market for decades. The main advantage of UHF RFID systems over the former techniques is their longer read-range and faster reading procedure, extending their range of applications dramatically. As these UHF RFID systems developed, their limitations and disadvantages became known. In response, other techniques including Near-Field Communication RFIDs and chipless RFIDs have been proposed. First, we will review the basic history of automatic contactless identification techniques, then go through some of the important techniques in more detail. The main purpose of this chapter is to give a brief overview of the important automatic contactless identification techniques.

1.1.1 Punched Card

Although punched cards might not be technically considered a contactless identification technique, we can categorize them within the automatic identification methodology. The history of punched cards goes back to the early seventeenth century when a textile worker named Basile Bouchon invented a punched paper tape to control textile looms (Fig. 1.1). His assistant, Jean-Baptiste Falcon, completed his invention 3 years later, although the machine was never mass-produced [1]. Jacques Vaucanson made a more complete version of this machine in 1745, and finally it was mass-produced as the Jacquard loom in 1805.

R. Rezaiesarlak, M. Manteghi, *Chipless RFID*, DOI 10.1007/978-3-319-10169-9_1

Fig. 1.1 Basile Bouchon's punched paper tape system on display at the Musée des Arts et Métiers, Paris

In 1892, Franco-Italian organ maker Anselmo Gavioli invented the book music medium, which enabled the storage of larger repertoires of music than ever before on thick folding cardboard "books". The books, shown in Fig. 1.2, are fed through the machine to control the organ, playing the song. These types of automatic fairground organs are still in use today.

Punched-card methodologies and techniques took a long evolutionary path through the work of different inventors, taking their most powerful step in finally becoming IBM punched cards, shown in Fig. 1.3, which were used for early computer programming. An IBM punched card is essentially a piece of stiff paper that contains stored information as the presence or absence of holes in predefined positions.

1.1.2 Barcode

The encoding and recording of information using different characters or symbols has a very long history. The major advantage of barcodes over former methods of recording coded information is the ability for machines to read a barcode automatically. This property makes the barcode the first automatic contactless identification technique.

"Classifying apparatus and method" is the name that Norman J. Woodland and Bernard Silver, students of Drexel Institute of Technology in Philadelphia, chose

Fig. 1.2 The Gavioli 87 Keyless Military Band Organ uses bursts of air to read the holes, instead of metal keys

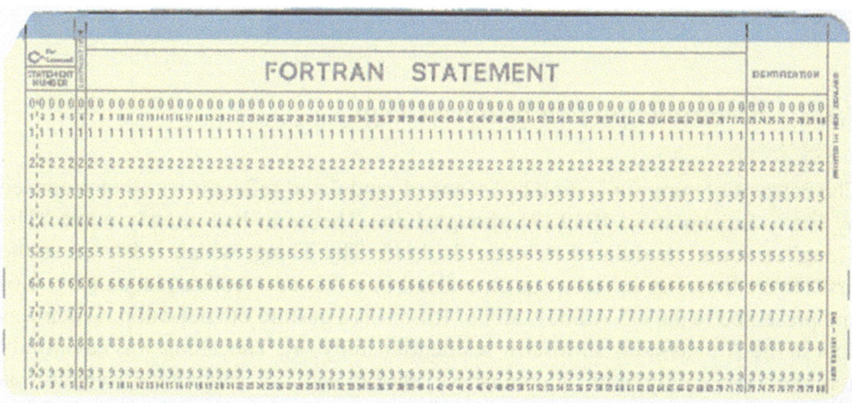

Fig. 1.3 IBM punched card, one of the first methods for computer programming

for their 1952 patent [2] on what is currently referred to as the barcode, Fig. 1.4. The motivation for this effort was a request from the president of a food chain, Food Fair, for a system to read product information automatically. A simple student project evolved into the most commonly used item tracking and automatic identification method of all time. The main advantage of barcodes is their low cost of production and ease of use. However, they suffer from limitations in the amount of information they can carry and range for their line-of-sight reading mechanism.

Fig. 1.4 Barcode

Fig. 1.5 Square barcodes can record larger amounts of information

Although there are a few new generations of square barcodes (Fig. 1.5) that can increase the amount of encoded information, they still suffer from the short line-of-sight read-range.

1.1.3 Electromagnetically-Based Tags

The concept of communication by means of modulated reflector was first proposed by Harry Stockman in 1949 [3], though Léon Theremin had already made "The Thing", a sound modulated reflector, in 1945. The Thing, which is known as the first covert listening device, was embedded in a carved wooden plaque (Fig. 1.6) and presented to the US Ambassador to Russia as a gesture of friendship, which was kept in the Ambassador's Moscow residential study [4]. The Thing was essentially a microwave cavity with a conductive elastic wall, or membrane, which was sensitive to variations in air pressure. The cavity-microphone was excited with a probe connected to a monopole antenna tuned to the resonance frequency of the cavity-microphone, 330 MHz. This frequency was selected to optimize The Thing's performance for the average frequency range of a male voice. The spying agency used to transmit a continuous wave (CW) at the cavity's resonance frequency. The time-varying wall frequency-modulated the stored energy in the cavity, which was then re-radiated by the same antenna, revealing the speech information near the device. This Trojan horse recorded and transmitted discussions in the US Ambassador to Russia's personal study for years. The theory of time-varying antennas is well-studied in [5–7].

Fig. 1.6 The Thing, or the Great Seal bug, was designed by Léon Theremin in 1945

1.1.3.1 Near-Field Magnetic Coupling RFID Tags

The first generation of commercialized RFIDs were designed for short distances to reduce the complexity and cost of fabrication of both tags and readers. Due to the short distance, there was no need for far-field radiation; the electromagnetic energy was transferred by means of reactive near-field coupling. Charles Walton filed his first patent of ten on a multi-resonator system in 1973. The system was comprised of an RFID tag and a sweeping oscillator, coupled to the tag's inductor, as the reader [8]. In fact, the first commercialized RFID tag, designed by Walton, did not have any electronic circuitry at all. Rather, it only consisted of multiple LC resonators; therefore, chipless RFID tags predate the conventional RFIDs.

Charles Walton filed another patent in 1984 [9] on a near-field RFID tag using electronics to improve the coding and decoding mechanism. There are different types of near-field RFID tags operating at various frequency bands, which will be discussed briefly in this book.

Near-field magnetic coupling tags, shown in Fig. 1.7, were overshadowed by UHF far-field tags due to the longer read-range and higher reading speed of the UHF tags. Then again, after a slow developmental period, near-field magnetic coupling tags came back as Near-Field Communication (NFC) RFID tags, and are currently the dominant technology in the RFID industry. Most of the modern smartphones are equipped with NFC devices capable of reading and writing NFC tags.

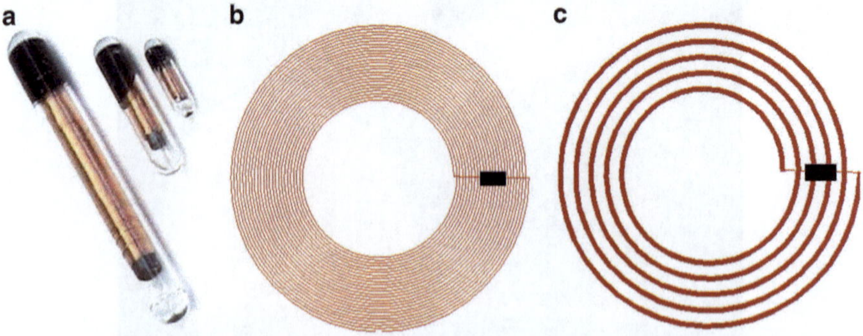

Fig. 1.7 Magnetic coupling RFID tags (**a**) ferrite core low frequency RFID tag. (**b**) Air-core low frequency RFID tag. (**c**) Air-core high frequency RFID tag

1.1.3.2 UHF RFID Tag

The first paper on UHF RFID was published in 1975 by Koelle and his colleagues at Sandia National Laboratories in Albuquerque, on producing a passive identifying system operating at UHF frequencies [10]. An overview paper on RFID was published by Rao in 1999 [11], followed by a presentation by Philips Semiconductor in 2002 [12] reporting a new RFID tag product that could be read at 3.3 m distance. The European Commission supported a research program on passive long-distance multiple-access UHF RFID systems [13] in 2001 which resulted in a UHF RFID tag designed by Karthaus and Fischer [14]. This marked the beginning of the race to design UHF RFID tags with lower price and higher sensitivity.

1.1.3.3 Chipless RFID Tag

Most of the former techniques of RFID tagging employed some type of time-variant loading or scatterer modulation to incorporate information into the backscattered field [15–20]. This backscattering modulation requires a dynamic scatterer; either the scatterer physically moves (e.g., The Thing, Identification Friend or Foe, IFF) or electronically changes (e.g., near-field magnetic coupling RFID, UHF RFID) through impedance modulation of the antenna or the coupling inductor. In contrast, a chipless RFID system uses the natural resonance of a metallic structure as its identification. This technique relies on the inherent static information incorporated in the scatterer structure.

The most commonly used chipless RFID tag, capable of storing only one bit of information, was designed by Thomas Thompson in 1956 [21] as a system for preventing shoplifting from retail stores. Later in the 1960s, two companies Checkpoint and Sensormatic designed and commercialized their own systems for Electronic Article Surveillance, EAS (Fig. 1.8). The tag would activate an alarm system located in the gate of the store if it had not been properly deactivated by the cashier.

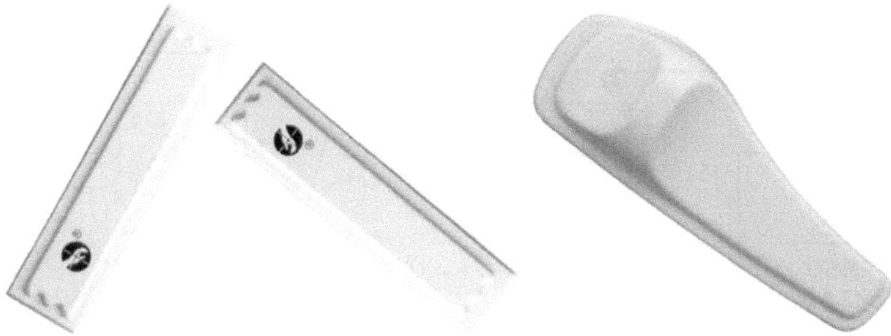

Fig. 1.8 Chipless RFID tags, capable of storing only one bit of information, are currently in use in most retail stores and libraries

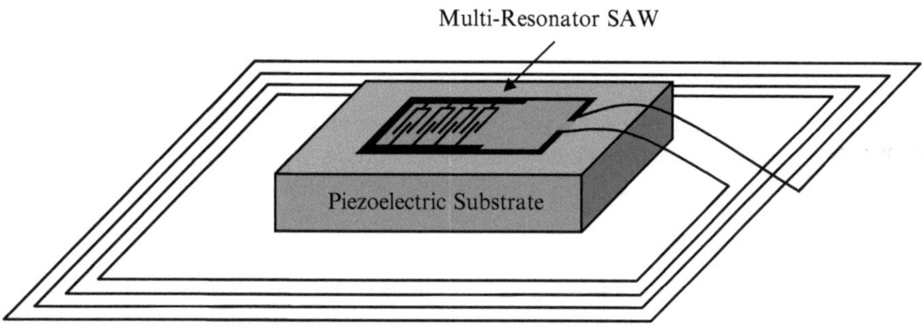

Fig. 1.9 Chipless RFID using Surface Acoustic Wave resonators [22]

The improved versions of these systems are still presently in use in most retail stores and libraries.

The first reported multi-bit chipless RFID tag communicating via Surface Acoustic Wave RFID was designed by Peter Harold Cole and Richard Vaughan in Australia in 1972 [22]. Their chipless RFID tag was interrogated by transmitting electromagnetic energy to the tag and receiving electromagnetic energy from the tag. The tag was printed on a piezo-electric substrate which served to receive electromagnetic energy, convert it to acoustic energy, store the converted energy for a suitable time, then reconvert the stored energy to electromagnetic energy and transmit that information to the receiver. The identification information was stored as different time delays in the substrate. They used the near-field coupling technique for their invention, consisting of a multi-notch SAW filter connected to a multi-turn inductor as the coupler (Fig. 1.9).

Other researchers have followed their work since 1996 to develop SAW-based chipless RFID with higher resolution and lower losses [23–26]. The main issue with SAW RFID is their ohmic loss as an inherent property of SAW technology.

Almost at the same time as the conception of the SAW RFID tag, Charles Walton designed and commercialized his chipless RFID tag in 1973 [8] using LC resonators as natural resonances (Fig. 1.10). He incorporated multiple LC

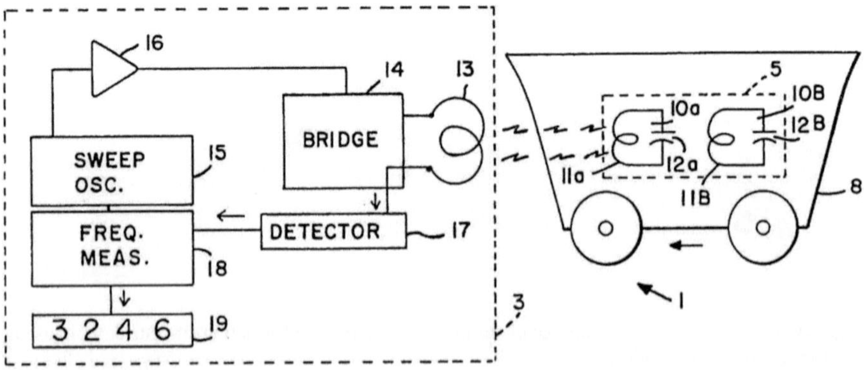

Fig. 1.10 Electronic identification and recognition system invented by Walton [8]. US patent number: US3752960 A 1973

resonators in the frequency range of 3–32 MHz as electronic keys for the Schlage Lock Company. The main issue with his technique was its low resolution and limited information that each key could carry. The other challenge was intrinsic, as the magnetic near-field requirement of his invention limited the read-range of the tag.

A thesis on low-cost electromagnetic tagging technology for the wireless identification, sensing, and tracking of objects was presented by Richard Ribon Fletcher at Massachusetts Institute of Technology in 1997 [27] which includes interesting information. For example, the concept of the chipless RFID tag as a sensor was proposed for the first time in this thesis.

The concept of a radio-frequency barcode using short-circuited dipoles with different lengths as resonators was proposed by Imad Jalaly in 2005 [28]. McVay presented his space-filling curve as a chipless RFID in 2006 [29]. Later in 2007, Majid Manteghi introduced a chipless RFID tag with multiple quarter-wavelength slots as high quality-factor resonators [30]. He mentioned the concept of natural resonances as the identification of chipless RFID tags in his paper with a sample three-bit chipless RFID tag. This tag can be printed with higher data density [31] using conductive ink with different patterns as logo or a cosmetic figure with a small size and minimum possible insertion loss on any product. Another research group in Australia adopted the concept of the SAW RFID tag by substituting the surface acoustic wave circuitry with printed-circuit resonators in addition to two orthogonal planar monopole antennas as receiver and transmitter in 2008 [32]. This design and SAW RFID share similar issues with high insertion loss and large size. Jalaly's work has been extended to various designs by Tedjini and his team [33] since 2010.

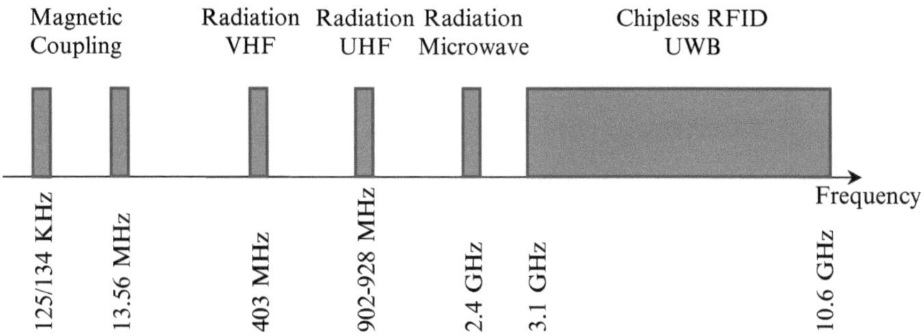

Fig. 1.11 Some of the frequency bands used in the US for RFID technology

1.2 NFC RFID System

There are different categorizations for RFID tags: inductive or radiative, active or passive, chipped or chipless. The inductive technique applies to RFID tags operating below 100 MHz. There are two different types of RFID technology in this category: Low-Frequency tags (operating at 125/134 kHz), and High-Frequency tags (operating at 13.56 MHz). Radiative tags, operating in the VHF range or above, will be discussed in the next sections. The electromagnetic connectivity is through the reactive near field for the magnetic coupling tags. An air-core transformer containing two coupled inductors, tuned to the operating frequency, provides the communication channel. Figure 1.11 shows the frequency bands commonly used by different RFID technologies.

Faraday introduced the concept of induction based on the electromotive force induced by a time-varying magnetic flux in a metallic structure in 1831. This concept was used by Nicholas Callan to invent the first transformer in 1836. A transformer consists of a minimum of two inductors that are magnetically coupled and whose magnetic fluxes are able to interact. Tesla announced his invention [34] using wireless power transfer for long distances and high power applications in 1891, although this invention has never been practically implemented the way Tesla envisioned. The idea of wireless power transfer was modified for low-power applications by Mario Cardullo and William Parks in 1973 [35] with their wireless transponder, which is the ancestor of current RFID transponders. It is worth reading their exact definition of their invention:

> A novel transponder apparatus and system is disclosed, the system being of the general type wherein a base station transmits an "interrogation" signal to a remote transponder, the transponder responding with an "answerback" transmission. The transponder includes a changeable or writable memory, and means responsive to the transmitted interrogation signal for processing the signal and for selectively writing data into or reading data out from the memory. The transponder then transmits an answerback signal from the data read-out from its internal memory, which signal may be interpreted at the base station. In the preferred inventive embodiment, the transponder generates its own operating power from the transmitted interrogation signal, such that the transponder apparatus is self-contained [35].

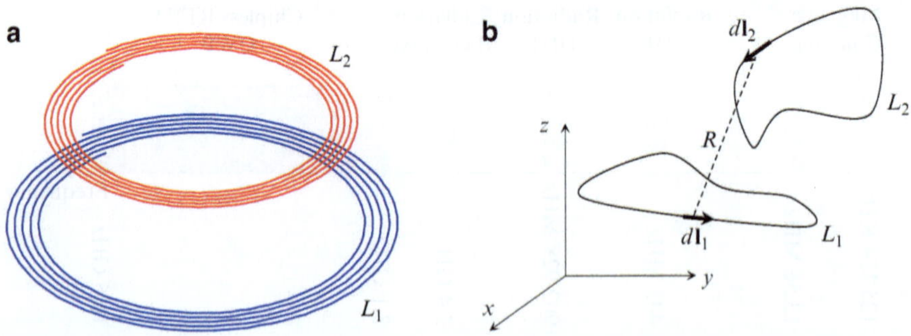

Fig. 1.12 (**a**) An air-core transformer contains two mutually coupled inductors. (**b**) Two coupled inductors with arbitrary shapes and locations. The coupling coefficient is a function of each inductor's topology, and their relative distance and orientation

By substituting "reader" for "base station" and "tag" for "transponder", we will have essentially the definition of a modern RFID system. They even proposed a communications protocol for their invention. These days, the EPC Gen II and ISO/IEC 14443 are being used for UHF and NFC tags, respectively, as improved versions of Cardolio and Parks's protocol.

The NFC RFID system relies on the mutual coupling between two tuned inductors. The time-variant magnetic flux generated by a current through the first inductor, L_1, which is connected to the reader, induces another current in the second inductor, L_2, which is connected to the tag (Fig. 1.12). Faraday's law explains the coupling mechanism, which is both a function of the inductors' topologies and their relative distance and orientation.

The required energy to power up the tag is transferred from the reader to the tag through the magnetic coupling between L_1 and L_2. At the same time, reader and tag communicate through their magnetic fields as well. A simple block diagram of the NFC RFID system is presented in Fig. 1.13.

To simplify the analysis, we use hard switches for both the tag and the reader. Figure 1.13a shows the power transfer state. Both switches are ON, and the power flows from reader to tag through their magnetic coupling.

$$\begin{cases} v_1 = L_1 \dfrac{di_1}{dt} - M \dfrac{di_2}{dt} \\ v_2 = -M \dfrac{di_1}{dt} + L_2 \dfrac{di_2}{dt} \end{cases} \tag{1.1}$$

M represents the mutual coupling between L_1 and L_2, which can be computed using Newmann's formula:

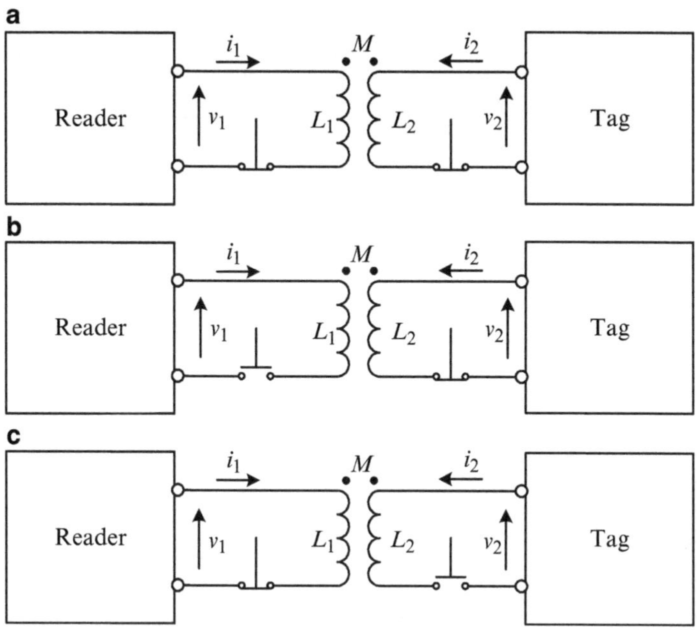

Fig. 1.13 Near-field RFID system including reader and tag. (**a**) Both inductors are in the circuit and the reader powers up the tag. (**b**) Reader transmits data by switching its inductor ON/OFF. (**c**) Tag responds back by switching its inductor ON/OFF

$$M = \frac{\mu}{4\pi} \oint_{L_1} \oint_{L_2} \frac{d\mathbf{l}_1 \cdot d\mathbf{l}_2}{R} \tag{1.2}$$

Here, $d\mathbf{l}_1$ and $d\mathbf{l}_2$ are the element currents on L_1 and L_2 respectively, shown in Fig. 1.13b, respectively. The induced current in L_2 generates a magnetic flux, which must oppose the original magnetic flux generated by i_1 based on Lenz's law. The induced current, i_2, will be rectified to provide the required power to turn on the logic circuitry in the tag. Then, the reader starts to switch its current, i_1, ON and OFF to send binary data to the tag (Fig. 1.13b). The demodulator in the tag circuitry (i.e., an envelope detector) detects the data, and the tag logic circuitry will respond through its switch (Fig. 1.7c). When the current in the tag is switched OFF ($i_2 = 0$), the voltage in the reader (v_1) will change to

$$v_1 = L_1 \frac{di_1}{dt} - M \frac{di_2}{dt} \overset{i_2=0}{\rightarrow} v_1 = L_1 \frac{di_1}{dt} \tag{1.3}$$

The demodulator at the reader is able to detect this time-varying signal and decode the tag information. Most of the modern NFC RFID systems follow the ISO/IEC 14443-A or -B [36] protocol for their communication. An overview of the architecture of an NFC tag is presented in Fig. 1.14. The essential components of the

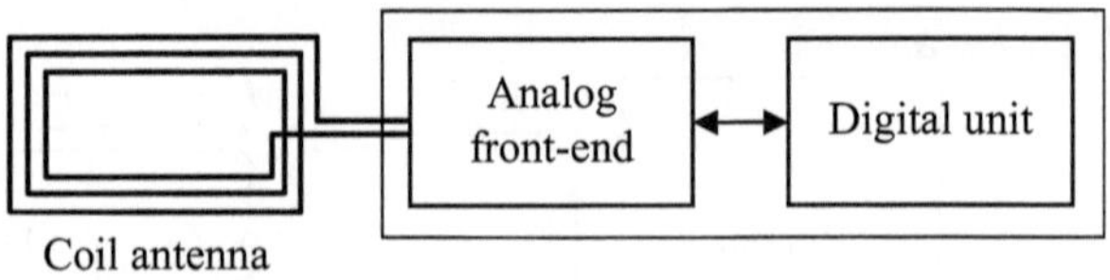

Fig. 1.14 Overview architecture of an NFC RFID tag

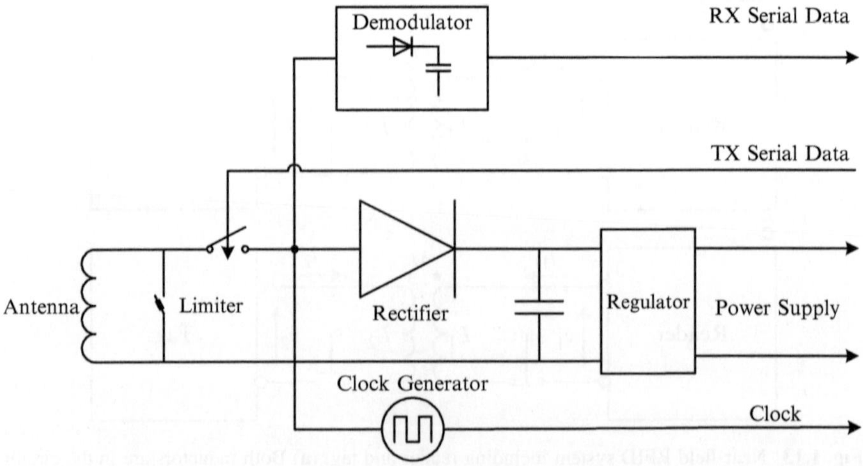

Fig. 1.15 Block diagram of the front-end of an NFC RFID tag

NFC tag are a distributed inductor as the coil antenna, analog front-end, and a digital logic unit.

The analog front-end is responsible for receiving data and power from the reader and transmitting data to the reader. The power chain begins with a limiter and continues with a rectifier and regulator in addition to storage capacitor. There will be a modulator and a demodulator in the analog front-end to interact with the digital unit. A simple block diagram for the analog front-end is presented in Fig. 1.15.

The digital unit contains a microcontroller in addition to I/O and different types of memory (e.g., EPROM, RAM, ROM). It is designed to receive and process serial data. It also has to handle the communications protocol and generate the required serial data to be transmitted to the reader.

1.3 UHF RFID Tag Design

Although modern designs for a UHF RFID tag include many different components which vary among manufacturers, the basic block diagram of a tag can be presented as in Fig. 1.16 [14]:

The essential modules of a UHF RFID are as follows: antenna, voltage multiplier, digital logic and memory, modulator, and demodulator.

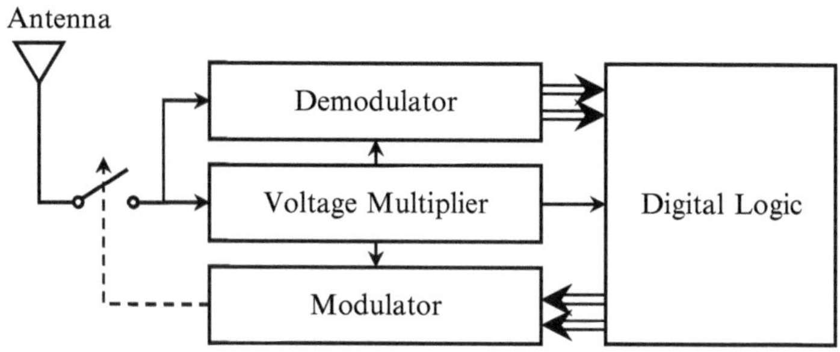

Fig. 1.16 Basic block diagram for a UHF RFID tag

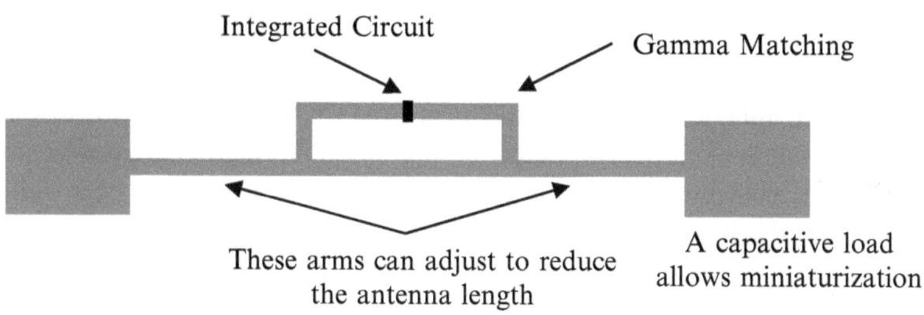

Fig. 1.17 Gamma-matched dipole as antenna for UHF RFID tag. To reduce the length, meandering arms can be used for the dipole antenna

1.3.1 Antenna

A gamma-matched dipole, shown in Fig. 1.17, has been used in most UHF RFID tags. Even so, magnetic antennas (slot antenna or short-ended parallel plate transmission line as a small loop) have been proposed for tags operating near metallic structures to overcome the image effect of the metallic structure.

The input impedance of the antenna is complex-conjugate matched to the input impedance of the silicon chip, $Z_c = Z_A^*$, to realize maximum power delivery to the tag circuitry. Different companies make chips with different input impedances which usually have a low real part and capacitive imaginary part ($Z_c = 27 - j201$ at 915 MHz for Alien Higgs-3). Based on EPC Gen2, one of the modulation schemes that can be used by tags to transmit data is ASK. An electronic switch or a voltage-controlled capacitor, a *varactor*, changes the antenna termination load to alter the radar cross section of the antenna. This way, the scattered field from the tag will vary by the impedance modulation and return as backscattered signal. The reader can detect the tag's antenna load variation and interpret that as received data (Fig. 1.18).

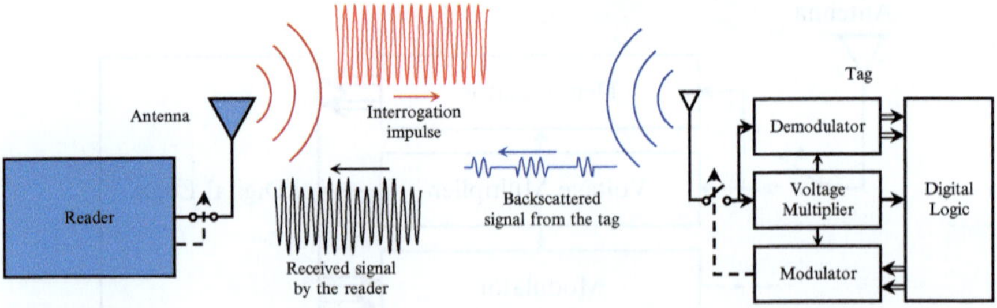

Fig. 1.18 RFID system. The reader is transmitting a CW signal and the tag is transmitting data by switching the antenna ON and OFF

1.3.2 Voltage Multiplier

The main disadvantage of using reactive near field for wireless power transfer in an NFC RFID system is its short range of operation, though high power-transfer efficiencies are possible within that range. In contrast to reactive near field, the radiative far field allows one to extend the UHF RFID range of operation far away from the reader antenna; however, the power-transfer efficiency decays dramatically over this distance. There are two constraints which limit the read range of UHF RFID tags: the maximum radiated power from the reader antenna, regulated by Federal Communications Commission (FCC), and the minimum required power for the monolithic chip of the RFID tag to turn on, which is limited by the chip technology. A simplified version of the Friis equation can be used to find an estimated read range:

$$R_{\max} = \frac{\lambda}{4\pi}\sqrt{\frac{P_{\text{reader}}G_{\text{reader}}G_{\text{tag}}}{P_{\min}}} \tag{1.4}$$

Here, λ is the wavelength, P_{reader} is the radiated power by the reader, G_{reader} is gain of the reader antenna, G_{tag} is the gain of the tag antenna, and $P_{\min}$ is minimum power required by the tag to turn on. Based on the FCC regulation reported in [37], the maximum equivalent Isotropically Radiated Power ($\text{EIRP} = P_{\text{reader}}\ G_{\text{reader}}$) for a reader at ISM 902–928 MHz band is 4 W. If we consider 0 dB gain for the tag antenna and $P_{\min} = -15\ dB_m$, the maximum read range will be $R_{\max} = 3$ m at 915 MHz while the tag is isolated in free space and there is no polarization mismatch. At the ideal matching case, one can compute the voltage at the chip ports ($Z_c = 27 - j201$) for $-15\ dB_m$ input power as $41\ mV$, which is not sufficient to power up any electronics in the tag circuitry. The classical way to overcome this issue is to use a diode-capacitor voltage multiplier. Figure 1.19 illustrates a full-wave rectifier voltage multiplier.

There have been various topologies proposed for voltage multipliers, with different pros and cons. The number of stages and type of diodes determines the multiplication constant.

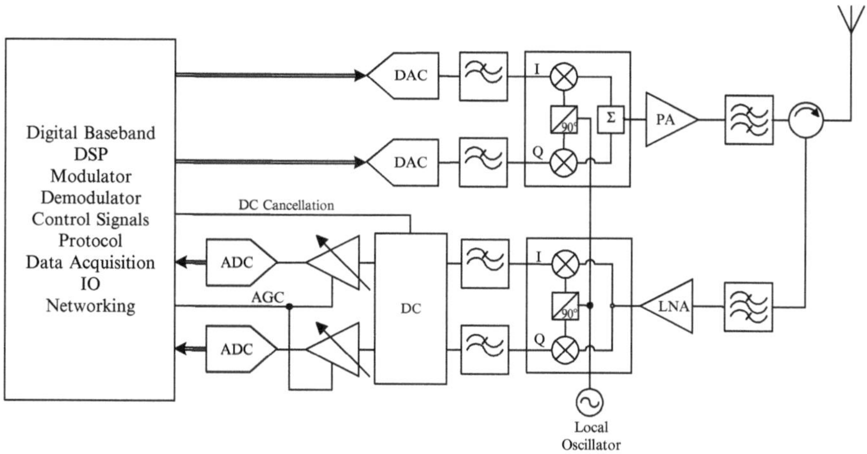

Fig. 1.21 Block diagram of a UHF RFID reader. This block diagram can be simplified using a variable-gain amplifier as ASK modulator

matching condition and 50 % backscattering efficiency in the tag structure while the transmitter antenna is radiating 1 W power. The minimum required power to turn ON the tag would be $-15\ dB_m$ which allows a 45 dB total link budget. The minimum received signal at the reader antenna for a reciprocal channel would be as follows: $-15\ -45\ -3 = -63\ dB_m$. The leakage from the transmitter to the receiver path ($30\ dB_m - 20\ dB_m = 10\ dB_m$) is still much stronger than the minimum received signal from the tag ($-63\ dB_m$). Therefore, the most effective architecture for the reader would be a direct down-conversion at the receiver side, which converts the CW signal, leaked from the transmitter, to a DC voltage. The next stage after the quadrature down-converter and the low-pass filters would be a DC cancellation circuitry to remove the down-converted leakage signal from the transmitter to the received signals.

DC cancellation is an important stage in the receiver architecture since the interference from the transmit signal could be 73 dB stronger than the received signal from the tag (worst-case scenario). There are many different innovative designs that have been proposed for DC cancellation. The block diagram presented in Fig. 1.21 shows the DC cancellation section, located before the baseband amplifier and controlled by the digital baseband circuitry. This way, the strong leaked signal will not go through the baseband amplifiers.

Variable-gain amplifiers and the AGC system are located right after the DC cancellation block. The simpler reader designs do not have any AGC systems; they take advantage of the ADC resolution to overcome the dynamic range of the received signal. The last stage before the digital baseband would be analog-to-digital converters.

1.4.2 Transmitter

Transmitter design is not as sophisticated as the receiver design in an RFID reader. There is a wide range of architectures proposed for the transmitter. A simple design consists of a variable-gain preamplifier connected to a power amplifier, followed by a bandpass filter. By altering the gain of the preamplifier between the minimum and maximum, one can generate a double-sideband amplitude shift-keying (ASK) signal. The proposed transmitter design in Fig. 1.21 allows one to generate any arbitrary single-sideband signal, which makes the design more flexible. This allows it to generate ASK or PSK signals to satisfy all Electronic Product Code (EPC) standards for transmitted signal.

1.4.3 Digital Baseband

As previously mentioned, baseband architecture without any intermediate frequency (IF) stages would be an adequate design for RFID reader. Therefore, all the analog-to-digital and digital-to-analog circuitries in the signal path operate at the baseband. The digital baseband usually consists of a processing unit in addition to different types of memory and I/O. Designs that are more sophisticated may use an FPGA for filtering and demodulation purposes as well. The digital baseband is responsible for modulating and demodulating the transmitted and received signals, respectively. It also handles the protocol and is responsible for data acquisition. Most of the modern readers are required to operate within a network of readers, controlled by a central unit.

1.5 Chipless RFID System

Instead of using a time-variant RCS by impedance modulation of the antenna like in conventional RFID tags, chipless RFID employs static information incorporated in the complex frequency domain of the RCS of the tag. The tag structure could be an arbitrary metallic body with some narrow-band natural resonant frequencies. The narrow-band resonances can be realized using simple resonant structures such as monopoles, dipoles, rings, half-wavelength slots, or quarter-wavelength slots. Although the frequency range of interest is assumed to be ultra-wideband (3.1–10.6 GHz), narrower bands at different frequency ranges can be used as well. The detection techniques provided in this book are for general applications and can be used for any type of chipless RFID tag. The overall system of a chipless RFID tag is shown in Fig. 1.22.

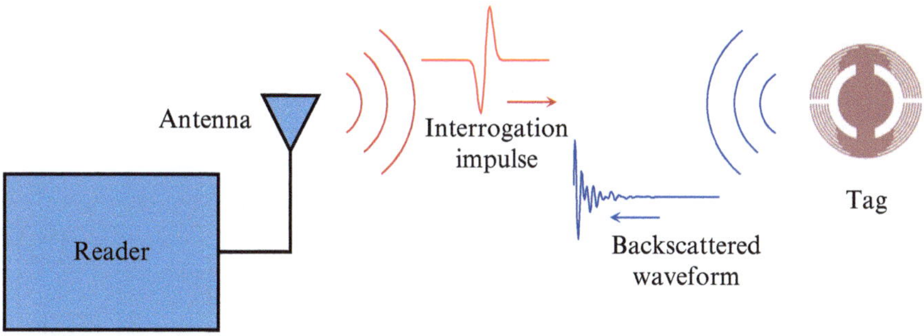

Fig. 1.22 Chipless RFID system including the reader and the chipless tag

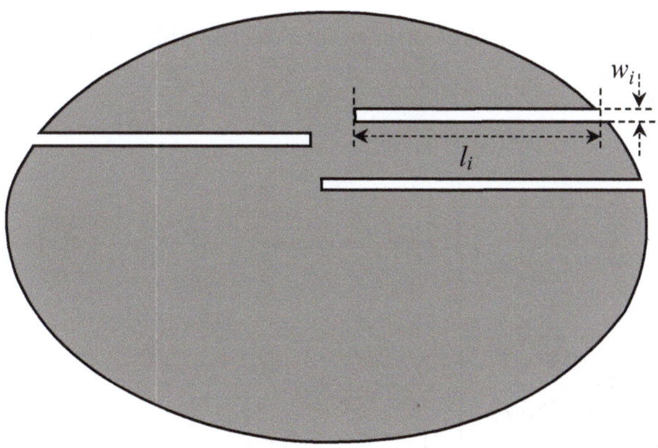

Fig. 1.23 A metallic patch with a few quarter-wavelength slots can be used as a chipless RFID tag

1.5.1 Chipless RFID Tag

A planar metallic patch with a few quarter-wavelength slots (Fig. 1.23) as identifications can be used as a simple structure for a chipless RFID tag [30].

The resonant frequency and quality factor of the *i*th quarter wavelength slot is determined by l_i and w_i. The design procedure and role of each component of the tag structure will be discussed in detail in subsequent chapters.

1.5.2 Chipless RFID Reader

Chipless RFID does not have any integrated logic circuitry to implement any communications protocol; therefore, the detection procedure has many extra challenges in comparison to the conventional UHF RFID tags. The reader should be able to extract the complex natural frequencies from the scattered signal. The signal will collide with signals from other scatterers or tags. This interference is called

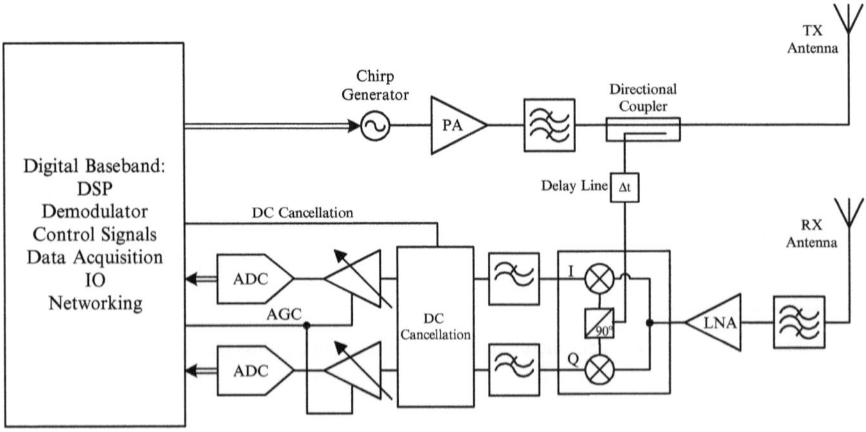

Fig. 1.24 Basic block diagram for a chipless RFID reader

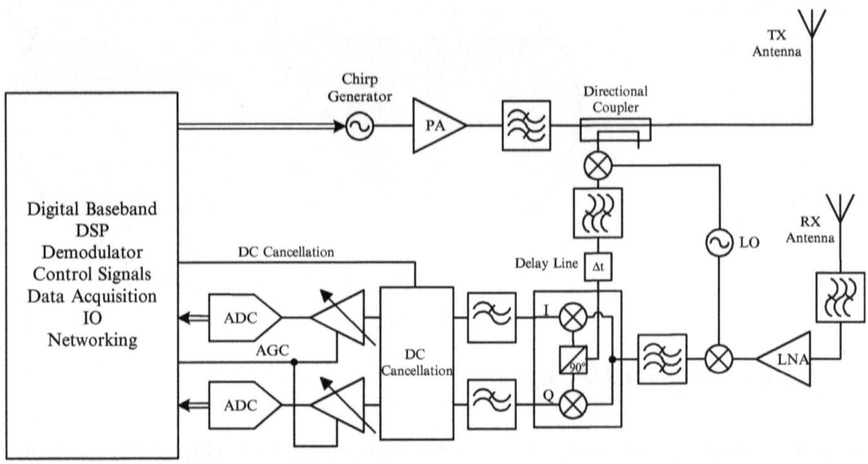

Fig. 1.25 Block diagram for a chipless RFID reader with an extra down conversion stage

'clutter', and the reader must detect the message signal through clutter. A basic architecture for a chipless RFID reader is presented in Fig. 1.24.

This architecture is a Frequency-Modulated Continuous Wave (FMCW) chirped radar, which has been used since 1960 [41]. The proposed UWB (3.1–10.6 GHz) reader takes advantage of a direct down-conversion in the receive path to overcome the transmitter interferences. Transmitter and receiver can either have two different UWB antennas or share one antenna in addition to a UWB circulator.

Figure 1.25 shows another architecture for the chipless RFID reader, which employs a local oscillator to down-convert the RF signals, simplifying the design of the quadrature demodulator. The digital baseband circuitry controls the LO frequency (which can be a sweeper as well to compress the RF bandwidth) and the chirp generator.

1.6 Book Outline

The contents of this book are presented in five chapters. In Chap. 1, a brief history of different identification techniques, including automatic contactless identification techniques, are presented here. Then, some of the basic architecture and methodology for near-field magnetic coupling RFID is discussed as one of the major players in current RFID technology. UHF RFID is discussed next. Then the basic architecture of the tag and reader is presented, in addition to some basic link budget calculation. The last section is devoted to chipless RFID, some of the methodology related to which is presented in this chapter.

A theoretical basis for the design and detection of chipless RFID tags is presented in Chap. 2. Some electromagnetics theorems are studied and employed for expanding the current distribution and equivalently, the radiated fields from the tag. First, the singularity expansion method (SEM) is discussed in more detail to demonstrate the physical concepts behind the complex natural resonances. Then, eigenvalue analysis followed by Characteristic Mode theory is presented in a generally rigorous formulation. These theorems will be widely used in the following chapters.

The theoretical concepts presented in Chap. 2 are used in designing chipless RFID tags in Chap. 3. Both SEM and characteristic mode theory are used in order to study the effects of the structural parameters on the radiation and resonant behavior of the tag. Different issues concerning with current designs are discussed based on both SEM and CM analysis, and a systematic design technique for improving the tag response is presented in this chapter.

In Chap. 4, the time and frequency characteristics of the scattered signal from chipless RFID tags are studied. Then, it will be shown that a time-frequency analysis is required in order to extract the required information of the tag in the reader. Various time-frequency techniques such as short-time Fourier transform (STFT), wavelet are studied in more detail. A time-frequency technique, called short-time matrix pencil method (STMPM) is introduced for the analysis of the scattered fields from the tag. We will discuss about the effects of various parameters of the STMPM on the time and frequency resolutions.

In the last chapter, an anti-collision algorithm is presented based on STMPM, by which the IDs and positions of different tags presented in the reader area are extracted. Some examples are presented in order to confirm the accuracy of the proposed technique.

References

1. Babbage C, Morrision P, Morrison E (1961) On the principles and development of the calculator and other seminal writings. Dover Publications, New York
2. Woodland NJ, Bernard S (1952) Classifying apparatus and method. Unite State Patent US 2612994 A

3. Stockman H (1948) Communication by means of reflected power. Proc IRE 36:1196–1204
4. Hatch B, ISECOM (Organization) (2008) Hacking exposed Linux: Linux security secrets & solutions, 3rd edn. McGraw-Hill, New York
5. Salehi M, Manteghi M (2014) Self-contained compact transmitter for high-rate data transmission. Electron Lett 50:316–318
6. Salehi M, Manteghi M, Suh S-Y, Sajuyigbe S, Skinner HG (2013) A wideband frequency-shift keying modulation technique using transient state of a small antenna. Prog Electromagn Res 143:421–445
7. Manteghi M (2010) Non-LTI systems, a new frontier in electromagnetics theory. In Antennas and Propagation Society International Symposium (APSURSI), 2010 IEEE, pp 1–4
8. Walton CA (1973) Electronic identification & recognition system. US3752960 A
9. Walton CA (1984) Electronic proximity identification and recognition system with isolated two-way coupling. US004600829
10. Koelle AR, Depp SW, Freyman RW (1975) Short-range radio-telemetry for electronic identification, using modulated RF backscatter. Proc IEEE 63:1260–1261
11. Rao KVS (1999) An overview of backscattered radio frequency identification system (RFID). In Microwave conference, 1999 Asia Pacific, vol 3, pp 746–749
12. CID ISO18000-6B presentation. Phillips Semiconductor, Cedar Rapids, IA, 2002
13. Annala A-L, Friedrich U (2001) Palomar—an European answer for passive UHF RFID applications? Available: http://citeseerx.ist.psu.edu/viewdoc/download?doi=10.1.1.201.7393&rep=rep1&type=pdf
14. Karthaus U, Fischer M (2003) Fully integrated passive UHF RFID transponder IC with 16.7-μW minimum RF input power. IEEE J Solid-State Circuits 38:1602–1608
15. Harrington RF (1963) Field measurements using active scatterers. IEEE Trans Microw Theory Tech 11:454–455
16. Richardson RM (1963) Remotely actuated radio frequency powered devices. US3098971 A
17. Harrington RF (1964) Theory of loaded scatterers. Proc Inst Elect Eng 111:617–623
18. Vogelman JH (1968) Passive data transmission technique utilizing radar echoes. US Patent 3391404
19. Rittenbach OE (1969) Communication by radar beams. US3460139 A
20. Vinding JP (1969) Interrogator-responder identification system. US3440633 A
21. Thompson TF (1956) Detecting means for stolen goods. US2774060 A
22. Cole PH, Vaughan R (1972) Electronic surveillance system. US Patent 3706094
23. Reindl L, Scholl G, Ostertag T, Ruppel CCW, Bulst WE, Seifert F (1996) SAW devices as wireless passive sensors. In Ultrasonics symposium, 1996. Proceedings, 1996 IEEE, vol 1 pp 363–367
24. Schmidt F, Scholl G (2001) Wireless SAW identification and sensor systems. In: Ruppel CCW, Fjeldly TA (eds) Advances in surface acoustic wave technology, systems and applications. World Scientific, Singapore, pp 277–325
25. Hartmann CS (2002) A global SAW ID tag with large data capacity. In Ultrasonics symposium, 2002. Proceedings. 2002 IEEE, vol 1, pp 65–69
26. Harma S, Arthur WG, Hartmann CS, Maev RG, Plessky VP (2008) Inline SAW RFID tag using time position and phase encoding. IEEE Trans Ultrason Ferroelectr Freq Control 55:1840–1846
27. Fletcher RR (1997) A low-cost electromagnetic tagging technology for wireless identification, sensing, and tracking of objects. MSc Thesis, School of Architecture and Planning, Massachusetts Institute of Technology
28. Jalaly I, Robertson ID (2005) RF barcodes using multiple frequency bands. In Microwave symposium digest, 2005 I.E. MTT-S International, 4pp
29. McVay J, Hoorfar A, Engheta N (2006) Space-filling curve RFID tags. In Radio and wireless symposium, 2006 IEEE, pp 199–202

30. Manteghi M, Rahmat-Samii Y (2007) Frequency notched UWB elliptical dipole tag with multi-bit data scattering properties. In Antennas and Propagation Society International Symposium, 2007 IEEE, San Diego, CA, pp 789–792
31. Rezaiesarlak R, Manteghi M (2014) Complex-natural-resonance-based design of chipless RFID tag for high-density data. IEEE Trans Antennas Propag 62:898–904
32. Preradovic S, Balbin I, Karmakar NC, Swiegers G (2008) A novel chipless RFID system based on planar multiresonators for barcode replacement. In 2008 I.E. international conference on RFID, pp 289–296
33. Deepu V, Vena A, Perret E, Tedjini S (2010) New RF identification technology for secure applications. In 2010 I.E. international conference on RFID-technology and applications (RFID-TA), pp 159–163
34. Cheney M (1981) Tesla, man out of time. Prentice-Hall, Englewood Cliffs, NJ
35. Cardullo M, Parks W (1973) Transponder apparatus and system. US3713148 A
36. I. International Organization for Standardization (2008) Identification cards—contactless integrated circuit cards—proximity cards. In ISO/IEC 14443
37. GS1 (2013) Regulatory status for using RFID in the EPC Gen 2 band (860 to 960 MHz) of the UHF spectrum. http://www.gs1.org/docs/epcglobal/UHF_Regulations.pdf
38. Mott DL (1970) On the radar cross section of a dipole. Proc IEEE 58:793–794
39. Feng X, Shaohui Q, Guoyu H (2008) RCS calculations of resonant dipole antennas with arbitrary loads based on the equivalent circuit method. In 8th International symposium on antennas, propagation and EM theory, 2008. ISAPE, pp 867–870
40. Griffin JD, Durgin GD (2009) Complete link budgets for backscatter-radio and RFID systems. IEEE Antennas Propag Mag 51:11–25
41. Klauder JR, Price AC, Darlington S, Albersheim WJ (1960) The theory and design of chirp radars. Bell Syst Tech J 39:745–808

Chapter 2
Mathematical Representation of Scattered Fields from Chipless RFID Tags

2.1 Introduction

As previously mentioned, a chipless RFID system is comprised of three basic components: reader, antenna, and chipless tag. The antenna illuminates the reader area and induces currents on the metallic tags. The induced currents re-radiate the scattered fields, which will be processed in the reader for decoding the IDs of the tags. The scattering phenomena is a sophisticated process, which can be described in a simple mathematical model. This mathematical model provides us insight of the electromagnetic behavior of the structure, which is useful in the design process of chipless RFID tags. The induced currents on the tag structure can be expanded in different ways. One method is to expand the induced currents versus the singularity poles of the tag, which is the basis of the singularity expansion method (SEM). In such a representation, the solution is expressed as a collection of poles, branch cuts, and an entire function in the complex frequency plane. In this chapter, after a comprehensive study of the SEM, the wavefront representation of the SEM is presented to describe the scattering mechanisms in the early-time and late-time modes. Altes' model is employed to describe the early-time response using the impulse responses of the scattering centers of the scatterer. Subsequently, the equivalent circuit of the scatterer is introduced based on the SEM representation of the fields. As an example, the current modes and associated radiated fields from a dipole antenna are studied.

On the other hand, the induced currents can be expanded versus the eigenmodes of the scatterer, which are found from an eigenvalue equation governing the structure under consideration. The singularity poles of the structure are related to the zeroes of the eigenvalues of the impedance matrix in the complex frequency plane. This relationship is computationally easy in extracting the complex natural resonances (CNRs) of the well-coordinated structures from the eigenmode expansion of the fields. As a simple case, the fields in a rectangular cavity resonator are represented versus the eigenmodes of its equivalent eigenvalue equation and the

R. Rezaiesarlak, M. Manteghi, *Chipless RFID*, DOI 10.1007/978-3-319-10169-9_2

relationship between the eigenvalues and singularity poles of the cavity is clearly shown in a closed-form representation.

As a weighted eigenvalue decomposition, the theory of characteristic modes is presented and effectively employed for analyzing the scattered fields from the chipless RFID tags. In this method, the current on the scatterer is mapped into a new set of characteristic modes, which are in-phase on the surface of the scatterer. Then, the currents are expanded versus these real-value characteristic modes. The characteristic-mode theory simplifies analysis of the induced currents, their related fields, and their behavior at different frequencies.

2.2 Singularity Expansion Method (SEM)

The concept of the singularity expansion method (SEM) was first introduced in 1971 by Carl Baum after observing the time-domain backscattered signal from missiles. By illuminating the target with a wideband pulse, it was observed that the time-domain returned signal from the missile includes some fast variations followed by damped sinusoidal signals. The history of SEM begins with a meeting at Northrop Corporate Laboratories in Pasadena, CA, in September 1971 [1]. After that meeting, Baum formulated the SEM in the Interaction note 88 [2]. He applied the proposed method to a perfectly conducting sphere and obtained the poles of the sphere in layers in the complex s-plane. After the advent of SEM, the proposed technique was broadly employed for the transient analysis of scatterers and antennas [2–10]. In [11–14], SEM was used to characterize the time and frequency-domain features of antennas and channels. Additionally, many researchers in the radar discipline employed SEM in identifying targets in military applications [15–18] and ground penetration radars [19–21]. Later on, the proposed technique was used in identifying and designing chipless RFID tags [22–24], breast cancer detection [25, 26], and monitoring the deployment of arterial stents implanted in blood vessels [27].

2.2.1 *Singularity Expansion Method in Circuit Theory*

The concept of the singularity expansion method (SEM) has been widely used in circuit theory for a long time [28]. As a simple case, an RLC-series resonator in Fig. 2.1a is considered, which is connected to a voltage source $v(t)$. The input impedance of a high-Q electrically small antenna can be modeled by this simple RLC circuit for a narrow frequency range [29–31]. The differential equation concerning the current in the circuit is

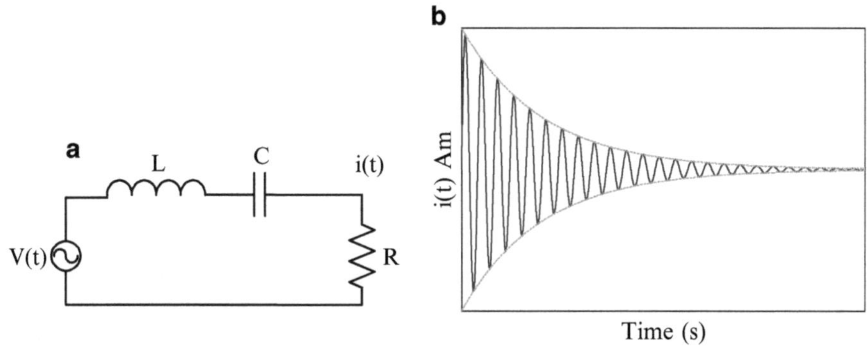

Fig. 2.1 (**a**) Series RLC circuit. (**b**) Impulse response of the circuit in time domain

$$\left(L\frac{d^2}{dt^2}+R\frac{d}{dt}+\frac{1}{C}\right)i(t)=\frac{dv(t)}{dt} \tag{2.1}$$

By applying the Laplace transform to (2.1) and assuming zero initial conditions for the voltage across the capacitor and current through the inductor, the current is written by

$$\begin{aligned} I(s) &= \frac{Cs}{\left(\frac{s}{\omega_0}\right)^2+\frac{1}{Q}\left(\frac{s}{\omega_0}\right)+1}\cdot V(s) \\ &= Y(s)\cdot V(s) \end{aligned} \tag{2.2}$$

Here, the radian resonant frequency, ω_0, and the quality factor at the resonant frequency, Q, are defined as

$$\omega_0=\frac{1}{\sqrt{LC}} \tag{2.3}$$

$$Q=\frac{1}{\omega_0 RC} \tag{2.4}$$

The admittance function, Y(s), can be expanded versus first-order poles of the circuit as

$$Y(s)=\frac{D}{s-s_1}+\frac{D^*}{s-s_1{}^*} \tag{2.5}$$

where $s_1=\alpha_1+j\omega_1$ is the complex natural resonance (CNR) of the circuit. The damping factor, resonant frequency, and D in (2.5) are given by

$$\alpha_1 = \frac{-\omega_0}{2Q} \tag{2.6}$$

$$\omega_1 = \frac{\omega_0}{2}\sqrt{4 - \frac{1}{Q^2}} \tag{2.7}$$

$$D = \frac{C}{2}\left(1 + j\frac{1 + \alpha_1}{\omega_1}\right) \tag{2.8}$$

For $Q \geq 0.5$, the circuit resonates at $f_1 = \omega_1/(2\pi)$. The residue of the pole in (2.2) is equal to $DV(s)$. Although the pole of the circuit is independent of the input voltage, the residue will change with variations in the input voltage. The impulse response of the circuit in the time domain is obtained as (2.9) by assuming $V(s) = 1$ in (2.2) and applying the inverse Laplace transform.

$$i(t) = |D|e^{-\alpha_1 t}\cos(\omega_1 t + \angle D) \tag{2.9}$$

Here, $i(t)$ is a damped sinusoidal current in the time domain as shown in Fig. 2.1b. Therefore, each complex natural resonance of a circuit acts as a damped sinusoidal signal in the time domain whose amplitude is related to the applied source. The above discussion can be easily generalized to multi-resonant circuits.

In a multi-resonant circuit, the input admittance of the circuit is expanded versus the singularity poles of the circuit as

$$Y(s) = F(s) + \sum_{n=-N}^{N} \frac{D_n}{s - s_n} \tag{2.10}$$

where N is the number of the CNRs, s_n, of the circuit and $F(s)$ is the entire function including the non-resonant part of the circuit [32]. In distributed circuits, such as cavity and transmission-line resonators where the number of resonances is infinity ($N \rightarrow \infty$), the entire function accelerates the convergence of the series [32].

2.2.2 Singularity Expansion Method in Transient Scattering

In order to study the scattering mechanism in chipless RFID systems, a 3-bit chipless RFID tag shown in Fig. 2.2 is illuminated by an incident field ($\mathbf{E}^{\mathbf{inc}}$, $\mathbf{H}^{\mathbf{inc}}$), launched from a transmitting antenna. The surrounding medium is assumed to be free space with permittivity ε_0 and permeability μ_0. The ID of the tag is set in the resonant frequencies of the structure, which can be embedded into some resonant-based circuitry on the tag. In practical applications, it is beneficial to design the tag on a metallic surface in order to maximize the radiation efficiency of the scatterer. Assuming the induced current on the tag as $\mathbf{J}$, the scattered field is obtained from either an electric-field integral equation (EFIE) or a magnetic-field

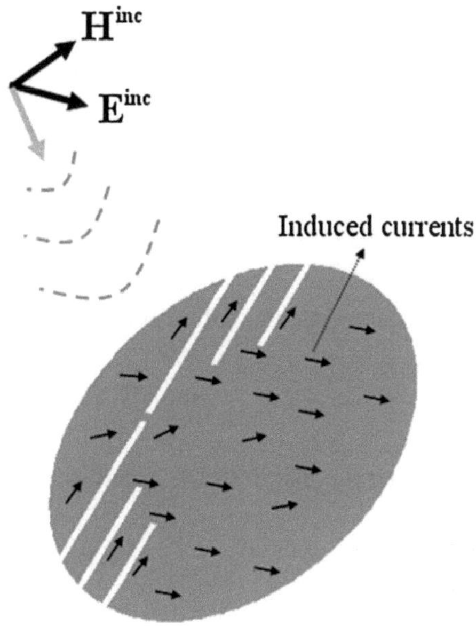

Fig. 2.2 3-bit tag illuminated by an incident plane wave

integral equation (MFIE) [33]. Here, the former case is considered for simplicity in formulations. Therefore, the scattered electric field is written in the Laplace domain as [34]

$$\mathbf{E}^{s}(\mathbf{r};s) = -\mu s \iint_{A} \left[\left(\overleftrightarrow{\mathbf{I}} - \frac{1}{k^2}\nabla\nabla \right) G_0\left(\mathbf{r},\mathbf{r}';s\right) \right] \cdot \mathbf{J}\left(\mathbf{r}';s\right) dS' \tag{2.11}$$

where $\overleftrightarrow{\mathbf{I}} = \hat{x}\hat{x} + \hat{y}\hat{y} + \hat{z}\hat{z}$, $s = \alpha + j\omega$ is the complex frequency, $k = s/c$ represents the propagation constant of the fields in the complex frequency domain, and A is the surface of the tag. The primed and unprimed coordinates represent the source and observation points, respectively. The quantity G_0 is the scalar Green's function in free space.

$$G_0\left(\mathbf{r},\mathbf{r}';s\right) = \frac{e^{-jk|\mathbf{r}-\mathbf{r}'|}}{4\pi|\mathbf{r}-\mathbf{r}'|} \tag{2.12}$$

G_0 satisfies the Sommerfeld radiation condition as

$$\lim_{r\to\infty} r\left(\frac{\partial}{\partial r} + jk\right) G_0\left(\mathbf{r},\mathbf{r}';s\right) = 0 \tag{2.13}$$

The scattered field in (2.11) can be written as the inner product of the dyadic Green's function and current distribution on the structure as

$$\mathbf{E}^{s}(\mathbf{r};s) = \left\langle \overleftrightarrow{\mathbf{G}}\left(\mathbf{r},\mathbf{r}';s\right),\mathbf{J}\left(\mathbf{r}';s\right)\right\rangle_{r'} \tag{2.14}$$

where the dyadic Green's function is defined as

$$\overleftrightarrow{\mathbf{G}}\left(\mathbf{r},\mathbf{r}';s\right) = -\mu s\left(\overleftrightarrow{\mathbf{I}} - \frac{1}{k^2}\nabla\nabla\right)G_0\left(\mathbf{r},\mathbf{r}';s\right) \tag{2.15}$$

The associated radiation condition for $\overleftrightarrow{\mathbf{G}}$ is

$$\lim_{r\to\infty} r(\nabla\times + jk\hat{r}\times)\overleftrightarrow{\mathbf{G}}\left(\mathbf{r},\mathbf{r}';s\right) = \mathbf{0} \tag{2.16}$$

and inner product in (2.14) is defined by

$$\langle \mathbf{A},\mathbf{B}\rangle_a = \int \mathbf{A}\cdot\mathbf{B}\ da \tag{2.17}$$

Assuming the tag is a perfect electric conductor (PEC), the boundary condition on the tag surface is given by

$$\hat{t}.\left(\mathbf{E}^{s}(\mathbf{r};s) + \mathbf{E}^{\text{inc}}(\mathbf{r};s)\right) = \mathbf{0} \qquad \forall\mathbf{r}\in A \tag{2.18}$$

where $\hat{t}$ denotes the unit vector tangential to the tag surface. As a result, the electric-field integral equation (EFIE) is written by

$$\left\langle \overleftrightarrow{\mathbf{G}}\left(\mathbf{r},\mathbf{r}'\right),\mathbf{J}\left(\mathbf{r}'\right)\right\rangle_{\mathbf{r}'} = -\mathbf{E}_{\text{t}}^{\text{inc}}(\mathbf{r}) \qquad \forall\mathbf{r}\in A \tag{2.19}$$

Subscript t in (2.19) indicates the tangential components of the fields on the tag surface. The method of moment (MoM) can be used to solve equation (2.19). By discretizing the surface of the tag into N isolated meshes and applying Galerkin's technique, one can write

$$\Gamma_{\text{mn}}\cdot \text{J}_{\text{n}} = \text{I}_{\text{n}} \tag{2.20}$$

The matrix equation in (2.20) should be in some sense an accurate representation of the integral equation in (2.19). One important criterion of such accuracy is the convergence of the solution obtained from (2.19) to the real current distribution as $N\to\infty$ [33]. The current distribution on the tag is obtained from

$$\text{J}_{\text{n}} = {\Gamma_{\text{mn}}}^{-1}\cdot \text{I}_{\text{n}} \tag{2.21}$$

According to (2.21), the singularity poles of the tag are the zeroes of the determinant of the coefficient matrix Γ as

$$\det(\Gamma(s_{\mathrm{k}})) = 0 \qquad k = 1,\ 2,\ 3\ldots \tag{2.22}$$

These singularity poles are the CNRs of the tag at which the current distribution on the tag shows damped oscillating behavior after the incident source field crosses through the tag. The basis of the SEM is that the current distribution is assumed to be an analytic function in the complex s-plane, except at CNRs such as

$$\mathbf{J}(\mathbf{r};s) = \sum_{n=-\infty}^{+\infty} \frac{\mathbf{a}_{\mathrm{n}}(\mathbf{r};s)}{s - s_{\mathrm{n}}} + \mathbf{J}_{\mathrm{e}}(\mathbf{r};s) \tag{2.23}$$

where $s_n = \alpha_n + j\omega_n$ is the nth CNR of the tag. Since the time-domain response is a real-valued signal, then

$$\begin{aligned} s_{\mathrm{n}} &= s_{\text{-n}}{}^{*} \\ \mathbf{a}_{\mathrm{n}}(\mathbf{r};s^{*}) &= [\mathbf{a}_{\mathrm{n}}(\mathbf{r};s)]^{*} \\ \mathbf{J}_{\mathrm{e}}(\mathbf{r};s^{*}) &= [\mathbf{J}_{\mathrm{e}}(\mathbf{r};s)]^{*} \end{aligned} \tag{2.24}$$

Equation (2.23) needs some more interpretation. According to Mittag-Leffler's theorem, an entire function in the s-plane is required for each pole in the infinite series to guarantee the convergence of the series [33, 35]. This entire function is represented by $\mathbf{J}_{\mathbf{e}}(\mathbf{r};\ \mathrm{s})$ in (2.23). The other important part of the series is the weighting function $\mathbf{a}_{\mathrm{n}}(\mathbf{r};\ s)$, which is assumed separable in the spectral-spatial form of

$$\mathbf{a}_{\mathrm{n}}(\mathbf{r};s) = R_{\mathrm{n}}(s)\mathbf{J}_{\mathrm{n}}(\mathbf{r}) \tag{2.25}$$

Here, $\mathbf{J}_{\mathrm{n}}(\mathbf{r})$ is the natural mode of the tag at the nth resonant frequency, and $R_n(s)$ is the corresponding frequency-dependent residue of the pole. By inserting (2.25) in (2.23), the current distribution close to s_{n} is written by

$$\mathbf{J}(\mathbf{r};s) = \frac{R_{\mathrm{n}}(s)\mathbf{J}_{\mathrm{n}}(\mathbf{r})}{s - s_{\mathrm{n}}} + \mathbf{J}_{e}(\mathbf{r};s) \tag{2.26}$$

By expanding $\overleftrightarrow{\mathbf{G}}$ and the incident source field, $\mathbf{E}^{\mathrm{inc}}$, in a power series around $s = s_{\mathrm{n}}$ as

$$\overleftrightarrow{\mathbf{G}}\left(\mathbf{r},\mathbf{r}';s\right) = \sum_{m=0}^{\infty} \frac{1}{m!} \left.\frac{\partial^{m}\overleftrightarrow{\mathbf{G}}(\mathbf{r},\mathbf{r}';s)}{\partial s^{m}}\right|_{s=s_n} (s - s_n)^{m} \tag{2.27}$$

$$\mathbf{E}_{\mathrm{t}}^{\mathrm{inc}}(\mathbf{r};s)=\sum_{m=0}^{\infty}\frac{1}{m!}\frac{\partial^{m}\mathbf{E}_{\mathrm{t}}^{\mathrm{inc}}(\mathbf{r};s)}{\partial s^{m}}\bigg|_{s=s_{\mathrm{n}}}(s-s_{\mathrm{n}})^{m} \tag{2.28}$$

and inserting (2.27) and (2.28) in (2.19), one can write

$$\begin{aligned}&\left\langle \overleftrightarrow{\mathbf{G}}\left(\mathbf{r},\mathbf{r}';s_{\mathrm{n}}\right)+(s-s_{\mathrm{n}})\frac{\partial\overleftrightarrow{\mathbf{G}}\left(\mathbf{r},\mathbf{r}';s\right)}{\partial s}\bigg|_{s=s_{\mathrm{n}}}+\cdots,\frac{R_{\mathrm{n}}(s_{\mathrm{n}})\mathbf{J}_{\mathrm{n}}\left(\mathbf{r}'\right)}{s-s_{\mathrm{n}}}+\mathbf{J}_{e}\left(\mathbf{r}';s\right)\right\rangle_{\mathbf{r}'}\\&=-\mathbf{E}_{\mathrm{t}}^{\mathrm{inc}}(\mathbf{r};s_{\mathrm{n}})-(s-s_{\mathrm{n}})\frac{\partial\mathbf{E}_{\mathrm{t}}^{\mathrm{inc}}(\mathbf{r};s)}{\partial s}\bigg|_{s=s_{\mathrm{n}}}-\cdots\end{aligned} \tag{2.29}$$

By balancing the two sides of (2.29) according to powers of $(s-s_n)$, some important expressions are obtained. The coefficient of the $(s-s_{\mathrm{n}})^{-1}$ term at $s=s_{\mathrm{n}}$ gives

$$\left\langle \overleftrightarrow{\mathbf{G}}\left(\mathbf{r},\mathbf{r}';s_{\mathrm{n}}\right),\mathbf{J}_{\mathrm{n}}\left(\mathbf{r}'\right)\right\rangle_{\mathbf{r}'}=\mathbf{0} \tag{2.30}$$

Equation (2.30) provides some important features of the CNRs and corresponding natural modes. By converting (2.30) to matrix form, it is seen that the determinant of the coefficient matrix should be zero at CNRs in order to have nontrivial solutions. As another significant point, these poles are completely dependent upon the dyadic Green's function of the structure and as (2.30) illustrates, they are source-free and aspect-independent parameters of the tag. This is the reason that these parameters are often used in identification applications. For each CNR, s_n, there is a nontrivial natural mode, $\mathbf{J}_n(\mathbf{r})$, which is the solution of (2.30). Corresponding to (2.30), one can define the coupling factors as the solutions to the following homogenous equation

$$\left\langle \mathbf{M}_{\mathrm{n}}\left(\mathbf{r}'\right),\overleftrightarrow{\mathbf{G}}\left(\mathbf{r},\mathbf{r}';s_{\mathrm{n}}\right)\right\rangle_{\mathbf{r}'}=\mathbf{0} \tag{2.31}$$

By equating the coefficients of $(s-s_{\mathrm{n}})^{0}$ in both sides of (2.29), one has

$$\begin{aligned}&\left\langle \overleftrightarrow{\mathbf{G}}\left(\mathbf{r},\mathbf{r}';s_{\mathrm{n}}\right),\mathbf{J}_{e}\left(\mathbf{r}';s\right)\right\rangle_{\mathbf{r}'}+R_{\mathrm{n}}(s_{\mathrm{n}})\left\langle\frac{\partial\overleftrightarrow{\mathbf{G}}\left(\mathbf{r},\mathbf{r}';s\right)}{\partial s}\bigg|_{s=s_{\mathrm{n}}},\mathbf{J}_{\mathrm{n}}\left(\mathbf{r}'\right)\right\rangle_{\mathbf{r}'}\\&=-\mathbf{E}_{\mathrm{t}}^{\mathrm{inc}}(\mathbf{r};s_{\mathrm{n}})\end{aligned} \tag{2.32}$$

The inner products in the left-hand side of (2.32) are performed on the $\mathbf{r}'$ parameter. Thus, both sides of the equation are functions of $\mathbf{r}$. By taking the inner products of the two sides of (2.32) by $\mathbf{M}_{\mathbf{n}}(\mathbf{r})$, the coupling coefficients can be found at resonant frequencies as

$$R_{\mathrm{n}}(s_{\mathrm{n}}) = -\frac{\left\langle \mathbf{M}_{\mathrm{n}}(\mathbf{r}), \mathbf{E}_{\mathrm{t}}^{\mathrm{inc}}(\mathbf{r}; s_{\mathrm{n}}) \right\rangle_{r}}{\left\langle \mathbf{M}_{\mathrm{n}}(\mathbf{r}), \left\langle \left. \frac{\partial \overleftrightarrow{\mathbf{G}}(\mathbf{r}, \mathbf{r}'; s)}{\partial s} \right|_{s=s_{\mathrm{n}}}, \mathbf{J}_{\mathrm{n}}(\mathbf{r}') \right\rangle_{\mathbf{r}'} \right\rangle_{\mathbf{r}}} \tag{2.33}$$

For electric-field integral equations (EFIE), where symmetric matrices are encountered, the coupling vectors and natural mode vectors are the same [33], so that (2.33) is written by

$$R_{\mathrm{n}}(s_{\mathrm{n}}) = -\frac{\left\langle \mathbf{J}_{\mathrm{n}}(\mathbf{r}), \mathbf{E}_{\mathrm{t}}^{\mathrm{inc}}(\mathbf{r}; s_{\mathrm{n}}) \right\rangle_{r}}{\left\langle \mathbf{J}_{\mathrm{n}}(\mathbf{r}), \left\langle \left. \frac{\partial \overleftrightarrow{\mathbf{G}}(\mathbf{r}, \mathbf{r}'; s)}{\partial s} \right|_{s=s_{\mathrm{n}}}, \mathbf{J}_{\mathrm{n}}(\mathbf{r}') \right\rangle_{\mathbf{r}'} \right\rangle_{\mathbf{r}}} \tag{2.34}$$

It is seen in (2.34) that the coupling coefficients at resonant frequencies depend on the incident electric field as well as the natural mode distribution at the corresponding resonant frequency. In the cases where $\langle \mathbf{J_n}(\mathbf{r}), \mathbf{E}_{\mathrm{t}}^{\mathrm{inc}}(\mathbf{r}; s_{\mathrm{n}}) \rangle_{\mathbf{r}} = 0$, the related mode will not be excited by the incident electric field. The coupling coefficients in (2.34) are just obtained at CNRs of the tag. There is no straightforward way to obtain the entire function added to the resonant response of the scatterer in (2.23). Mathematically, this is necessary to guarantee convergence of the series. However, more explanation is needed in order to understand the physical concepts behind the theory of SEM.

2.2.2.1 Coupling Coefficients and Turn-On Times

As the equation (2.34) shows, the coupling coefficients at the resonant frequencies of the structure depend on the natural modes, dyadic Green's function of the structure, and incident field at those frequencies. For other complex frequencies, s, different representations can be chosen as the coupling coefficient, which affects the entire function added to the series. In the late-time response of the scatterer, we have just the damped sinusoidals corresponding to the CNRs of the tag. Hence, the entire-function contribution comes into the early-time response, which rises and falls faster than the late-time signals. In order to cover other complex frequencies, different coupling coefficients have been introduced, where class 1 and class 2 representations are most common in literature [33]. For a class 1 representation, which is the simplest one, the coupling coefficients of the natural modes are defined as

$$R_{\mathrm{n}}^{(1)}(s) = e^{(s_{\mathrm{n}}-s)t_0} R_{\mathrm{n}}(s_{\mathrm{n}}) = -e^{(s_{\mathrm{n}}-s)t_0} \frac{\left\langle \mathbf{J}_{\mathrm{n}}(\mathbf{r}), \mathbf{E}_{\mathrm{t}}^{\mathrm{inc}}(\mathbf{r}; s_{\mathrm{n}}) \right\rangle_{\mathbf{r}}}{\left\langle \mathbf{J}_{\mathrm{n}}(\mathbf{r}), \left\langle \left. \frac{\partial \overleftrightarrow{\mathbf{G}}(\mathbf{r}, \mathbf{r}'; s)}{\partial s} \right|_{s=s_{\mathrm{n}}}, \mathbf{J}_{\mathrm{n}}(\mathbf{r}') \right\rangle_{\mathbf{r}'} \right\rangle_{\mathbf{r}}} \tag{2.35}$$

By inserting (2.35) in the series part in (2.23), the time-domain response is given by

$$\mathbf{j}(\mathbf{r}; t) = U(t - t_0) \mathrm{Re}\left[\sum_{n=1}^{\infty} R_{\mathrm{n}} \mathbf{j}_{\mathrm{n}}(\mathbf{r}) e^{s_{\mathrm{n}} t} \right] + \mathbf{j}_{\mathrm{e}}(\mathbf{r}; t) \tag{2.36}$$

where $U(.)$ is the Heaviside step function defined as

$$U(t - t_0) = \begin{cases} 1 & t \geq t_0 \\ 0 & t < t_0 \end{cases} \tag{2.37}$$

and the inverse Laplace transform is defined as

$$\mathbf{j}(\mathbf{r}; t) = \int_{Br} e^{st} \mathbf{J}(\mathbf{r}; s) ds \tag{2.38}$$

which causality ensured by having the Bromwich integration contour Br passing above all singularities in the s-plane. Turn-on time might be the time at which the incident wave is first applied anywhere on the tag. Although class 1 form of the coupling is more useful in the analytical-based formulation of SEM, it shows some convergence issues in earlier times of the response in numerical calculations [36]. For computational purposes, the class 2 form is more efficient. In this form, the frequency dependency of the coupling coefficients is held in the incident electric field as

$$R_{\mathrm{n}}^{(2)}(s) = -\frac{\left\langle \mathbf{J}_{\mathrm{n}}(\mathbf{r}), e^{(s_{\mathrm{n}}-s)t_0} \mathbf{E}_{\mathrm{t}}^{\mathrm{inc}}(\mathbf{r}; s) \right\rangle_{\mathbf{r}}}{\left\langle \mathbf{J}_{\mathrm{n}}(\mathbf{r}), \left\langle \left. \frac{\partial \overleftrightarrow{\mathbf{G}}(\mathbf{r}, \mathbf{r}'; s)}{\partial s} \right|_{s=s_{\mathrm{n}}}, \mathbf{J}_{\mathrm{n}}(\mathbf{r}') \right\rangle_{\mathbf{r}'} \right\rangle_{\mathbf{r}}} \tag{2.39}$$

The effect of these coupling coefficients on the current distribution can be better illustrated in the time domain. For more simplicity, the following incident electric field is considered.

$$\mathbf{E}_{\mathrm{t}}^{\mathrm{inc}}(\mathbf{r}; s) = \mathbf{E}_0 e^{-\frac{s}{c} \hat{\mathbf{r}} . \mathbf{r}} \tag{2.40}$$

where the vector $\hat{r}$ is the propagation vector, $\mathbf{E}_0$ includes the polarization vector and amplitude of the incident wave, and c is the speed of light in free space. By inserting

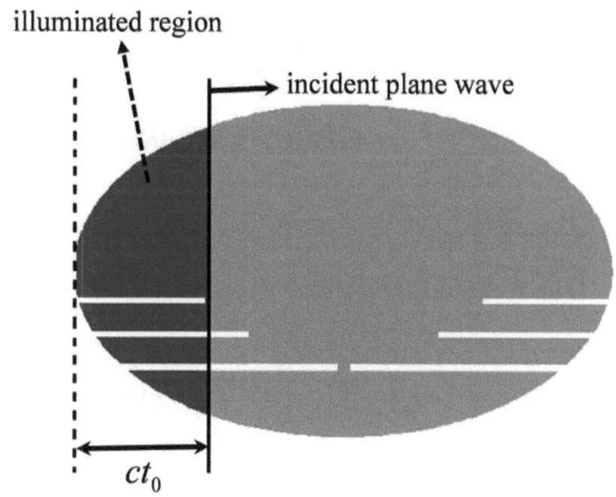

Fig. 2.3 3-bit tag illuminated partly by an incident plane wave

(2.39) and (2.40) in (2.23), the current distribution in the Laplace domain is written as

$$\mathbf{J}(\mathbf{r};s)=\sum_{n=-\infty}^{\infty}\frac{\left\langle \mathbf{J}_{\mathrm{n}}(\mathbf{r}),\mathbf{E}_0 e^{(s_{\mathrm{n}}-s)t_0-\frac{s}{c}\hat{\mathbf{r}}\cdot\mathbf{r}}\right\rangle_{\mathbf{r}}\mathbf{J}_{\mathrm{n}}(\mathbf{r})}{(s-s_{\mathrm{n}})\left\langle \mathbf{J}_{\mathrm{n}}(\mathbf{r}),\left\langle \left.\frac{\partial\overleftrightarrow{\mathbf{G}}(\mathbf{r},\mathbf{r}';s)}{\partial s}\right|_{s=s_{\mathrm{n}}},\mathbf{J}_{\mathrm{n}}(\mathbf{r}')\right\rangle_{\mathbf{r}'}\right\rangle_{\mathbf{r}}}+\mathbf{J}_{\mathrm{e}}(\mathbf{r};s) \quad (2.41)$$

By applying the inverse Laplace transform defined in (2.38) to (2.41), the current distribution in time domain is written as

$$\mathbf{j}(\mathbf{r};t)=2\sum_{n=1}\mathrm{Re}\left(\frac{\left\langle \mathbf{J}_{\mathrm{n}}(\mathbf{r}),\mathbf{E}_0 e^{-\frac{s_{\mathrm{n}}}{c}\hat{\mathbf{r}}\cdot\mathbf{r}}U\left(t-t_0-\frac{\hat{\mathbf{r}}\cdot\mathbf{r}}{c}\right)\right\rangle_{r}\mathbf{J}_{\mathrm{n}}(\mathbf{r})}{\left\langle \mathbf{J}_{\mathrm{n}}(\mathbf{r}),\left\langle \left.\frac{\partial\overleftrightarrow{\mathbf{G}}(\mathbf{r},\mathbf{r}';s)}{\partial s}\right|_{s=s_{\mathrm{n}}},\mathbf{J}_{\mathrm{n}}(\mathbf{r}')\right\rangle_{\mathbf{r}'}\right\rangle_{\mathbf{r}}}e^{s_{\mathrm{n}}t}\right)+\mathbf{j}_{\mathrm{e}}(\mathbf{r};t) \quad (2.42)$$

The convergence difficulties in the class 1 form of coupling coefficients are alleviated in the class 2 representation, where a time-varying region of integration covers that part of the object surface, which has already been illuminated by the incident field [36]. When the incident wave completely passes through the tag, both class 1 and class 2 representations are similar. For better illustration, Fig. 2.3 shows the region of integration at $t=t_0$ on the surface of the tag for class 2 coupling coefficients when the incident plane wave passes through a part of the tag.

2.2.2.2 Scattered Field from Chipless RFID Tags

By representing the current distribution on the tag as the summation over the natural modes in the late-time response accompanying an entire function as the early-time response, the scattered field is obtained from the integral equation in (2.14) as

$$\mathbf{E}^{s}(\mathbf{r};s) = \left\langle \overleftrightarrow{\mathbf{G}}(\mathbf{r},\mathbf{r}';s), \sum_{n} \frac{R_{n}(s)\mathbf{J}_{n}(\mathbf{r}')}{s-s_{n}} + \mathbf{J}_{e}(\mathbf{r}';s) \right\rangle_{\mathbf{r}'}$$
$$= \sum_{n} \frac{\left\langle \overleftrightarrow{\mathbf{G}}(\mathbf{r},\mathbf{r}';s), \mathbf{J}_{n}(s)R_{n}(s) \right\rangle_{\mathbf{r}'}}{s-s_{n}} + \left\langle \overleftrightarrow{\mathbf{G}}(\mathbf{r},\mathbf{r}';s), \mathbf{J}_{e}(\mathbf{r}';s) \right\rangle_{\mathbf{r}'} \tag{2.43}$$

The radiated field close to the nth CNR is written by

$$\mathbf{E}^{s}(\mathbf{r};s) = \left\langle -\mu s\left(\overleftrightarrow{\mathbf{I}} - \frac{1}{k^{2}}\nabla\nabla\right)G_{0}\left(\mathbf{r},\mathbf{r}';s\right), \frac{R_{n}(s)\mathbf{J}_{n}(\mathbf{r}')}{s-s_{n}} + \mathbf{J}_{e}\left(\mathbf{r}';s\right) \right\rangle_{\mathbf{r}'} \tag{2.44}$$

In the far field, the field in (2.44) can be approximated by

$$\mathbf{E}_{n}(\mathbf{r};s) = \left\langle -\mu s\left(\overleftrightarrow{\mathbf{I}} - \hat{r}\hat{r}\right)G_{0}\left(\mathbf{r},\mathbf{r}'\right), \left.\frac{R_{n}(s)\mathbf{J}_{n}(\mathbf{r}')}{s-s_{n}}\right|_{s=s_{n}} \right\rangle_{\mathbf{r}'}$$
$$- \left\langle \mu s\left(\overleftrightarrow{\mathbf{I}} - \hat{r}\hat{r}\right)G_{0}\left(\mathbf{r},\mathbf{r}'\right), \mathbf{J}_{e}\left(\mathbf{r}';s\right) \right\rangle_{\mathbf{r}'}$$
$$= \left\langle -\mu s\left(\overleftrightarrow{\mathbf{I}} - \hat{r}\hat{r}\right)\frac{e^{-jk(r-r'\cdot\hat{r})}}{4\pi r}, \left.\frac{R_{n}(s)\mathbf{J}_{n}(\mathbf{r}')}{s-s_{n}}\right|_{s=s_{n}} \right\rangle_{\mathbf{r}'}$$
$$- \left\langle \mu s\left(\overleftrightarrow{\mathbf{I}} - \hat{r}\hat{r}\right)\frac{e^{-jk(r-r'\cdot\hat{r})}}{4\pi r}, \mathbf{J}_{e}\left(\mathbf{r}';s\right) \right\rangle_{\mathbf{r}'} \tag{2.45}$$

As (2.45) expresses, in the far-field region, the scattered fields in the time domain are approximately proportional to the first derivative of the current distributions on the tag. In contrast, in the near field, the field distribution is mostly affected by the spatial derivatives of the currents. By applying the inverse Laplace transform to the scattered field in (2.43), the fields in the time domain are written as

$$e^{s}(\mathbf{r};t) = U(t-t_{0})\sum_{n} |R_{n}|e^{-\alpha_{n}t}\cos(\omega_{n}t + \phi_{n}) + e_{e}(\mathbf{r};t) \tag{2.46}$$

where the class 1 form of coupling coefficients is assumed in (2.46). According to (2.46), the scattered field from a tag is affected by two different phenomena. Early-time response, which is depicted by $e_{e}(r;t)$ in (2.46), is affected by the specular reflections from the scattering centers of the tag. The early-time response is followed by the series of damped sinusoidals with some weighting coefficients. The CNRs of the tag, shown by $s_{n} = \alpha_{n} + j\omega_{n}$, are aspect-independent parameters of the tag, not dependent on the direction, polarization, or distance to the tag's observation point. For this reason, they are well-suited to be used as the tag's ID. In Chap. 4, different techniques are proposed for extracting the CNRs from the scattered field.

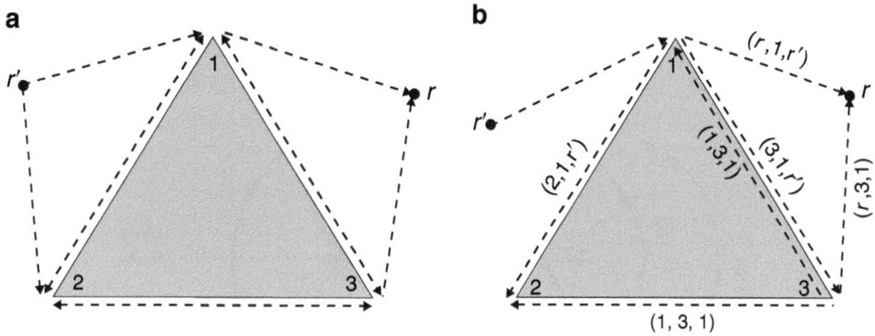

Fig. 2.4 (**a**) Scattering configuration. (**b**) Ray tracing from the source to observation point [37] with permission from IEEE

2.2.2.3 Wavefront Interpretation of the Singularity Expansion Method

As explained in previous sections, the CNRs and corresponding natural modes are aspect-independent parameters, which instead depend on the geometry and material of the tag. In order to ensure convergence of the series in the late time, an entire function was added to the current distribution responsible for the early-time variations. The shape of the entire function in (2.23) is dictated by the coupling coefficient representation chosen in the series. This part is called the removable early-time response [37]. In spite of the mathematical representation in (2.23), there is an intrinsic early-time response, which is related to the scattering process [37]. This part, with fast rise-and-fall variations in the time domain, cannot be described by the oscillatory waves from global CNRs of the tag. Rather, it can be explained by the progressive waves (wavefronts) with the help of the geometrical theory of diffraction (GTD) [37–39]. In the wavefront representation [37], the waves travel from source to tag. At the tag surface, they undergo an interactive process which appears to the observer as a series of wavefront arrivals corresponding to multiple passes through the tag and diffraction from those scattering centers. In [37], a hybrid method based on wavefront interpretation has been presented, which shows the global resonances to be a collective summation of multiple wavefront fields.

As an example, Fig. 2.4a shows a scatterer irradiated by an incident wavefront launched from a source located at r'. Three corners of the scatterer are considered as the scattering centers and the observation point is located at r. The situation shown in Fig. 2.4a is a bistatic situation, which can easily be adopted to the monostatic case by fixing the source and observation points at the same position. When a scatterer is illuminated by a pulse signal, first the specular reflections from the scattering centers are scattered around, emanating from the local current distribution on the discontinuities in the scattering surface. After the incident pulse passes through the scatterer, the local current distributions converge to the natural current modes of the scatterer, which are responsible for the late-time response. In other words, the interaction between local resonances in the early time produces the global complex natural

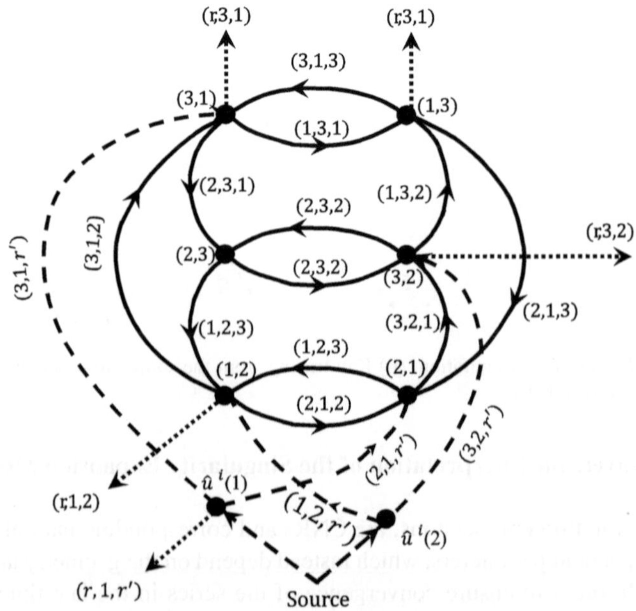

Fig. 2.5 Flow graph of the scatterer [37] with permission from IEEE

resonances of the scatterer. This behavior is illustrated in Fig. 2.4b. Two different rays are traced from the source point to the scatterer. Each path is illustrated by three components, (r_1, r_2, r_3), which express the field at the point r_1 caused by scattering at r_2 of a ray field with unit amplitude arriving from a point r_3.

In Fig. 2.4b, the incident ray hits the first scattering center. Part of it is received at the observation point via $(r, 1, r')$ and the other part flows to corner 3 through $(3, 1, r')$. Again, part of the ray in corner 3 hits the observation point through $(r, 3, 1)$ and the other part travels to the second corner by (2, 3, 1), and the third part reflects back to the first corner through (1, 3, 1). There is a local resonance in the path between corners 1 and 2. The other diffracted rays from the scattering centers are traced in the same way as mentioned before. This ray interaction can be easily traced with a signal flow graph, shown in Fig. 2.5. The rays scattered to the observation point are shown by dashed lines. For time-harmonic ray fields, a ray has the local plane-wave form of

$$S(r_1, r_2, r_3) \sim A(r_1, r_2, r_3)\exp[jk\psi(r_1, r_2, r_3)] \tag{2.47}$$

where A and $k\psi$ are slowly-varying amplitude and rapidly-varying phase functions, respectively. From the flow graph seen in Fig. 2.5, the ray paths from source to observation point can be divided into two groups. The first group, shown with a dashed line, contains contributions at r from direct interaction of the incident field with each of the scattering centers, which are shown by $\hat{u}^i(j), j = 1, 2, 3$ in Fig. 2.5.

The summation of these terms with weighting functions proportional to (2.47) includes the intrinsic early-time response of the scatterer caused by the scattering process. The second group accounts cumulatively for all the higher-order interactions schematized by the solid branches in the graph. This part of the response represents the SEM resonances of the scatterer. By returning to the general problem, scattered signal at the observation point can be written with Mason's formula as

$$u(r) = \sum_{n} u_{\mathrm{n}}^{\mathrm{d}} + \frac{1}{\Delta}\sum_{n} u_{\mathrm{n}}^{\mathrm{w}} \Delta_{\mathrm{n}} \tag{2.48}$$

where the first term is responsible for the early-time response and contains the direct interactions of the incident ray field with scattering centers. The second term, the late-time component of the response, represents the coupling between different local resonances on the surface of the scatterer. The quantity $u_{\mathrm{n}}^{\mathrm{d}}$ is the direct nth path from source to the observation point (shown by dashed lines in Fig. 2.5) and $u_{\mathrm{n}}^{\mathrm{w}}$ represents the wavefront contribution along paths going from source to observer via successive (more than one but without repetition) single scattering events. Δ_{n} shows the gain of the non-touching closed loops having no interaction with $u_{\mathrm{n}}^{\mathrm{w}}$, and Δ is the characteristic determinant of the graph. The zeroes of Δ are the CNRs of the scatterer. Although the wavefront representation of the scattering process cannot easily be employed for calculating the early-time and late-time responses of the tag, it provides a descriptive model for the scattering mechanism.

2.2.2.4 Altes' Model

Though the late-time response of the tag can be compactly cast into a series-form formulation, it is not as easy to predict the behavior of the early-time response. This is because it depends on the spatial variations of the scatterer and observation point. Based on (2.43), the early-time response is formulated by

$$\mathbf{E}_{\mathrm{e}}(r; s) = \left\langle -\mu s\left(\overleftrightarrow{\mathbf{I}} - \frac{1}{k^2}\nabla\nabla\right)G_0\left(\mathbf{r}, \mathbf{r}^{'}; s\right), \mathbf{J}_{\mathrm{e}}\left(\mathbf{r}^{'}; s\right)\right\rangle \tag{2.49}$$

where the first part of the dyadic Green's function is more pronounced in the far zone and the second term is dominant in the near zone of the scatterer. Because of the fast variations of the early-time currents, the scattered field includes pulse-shape impulses reflected from the scattering centers of the tag. Assuming a scatterer containing M scattering centers is illuminated by an incident plane wave with pulse function $p(t)$, the backscattered signal in the early time can be modeled as the summation of the delayed pulses from the scattering centers as [40]

Fig. 2.6 Transversal filter model of the early-time response [40]

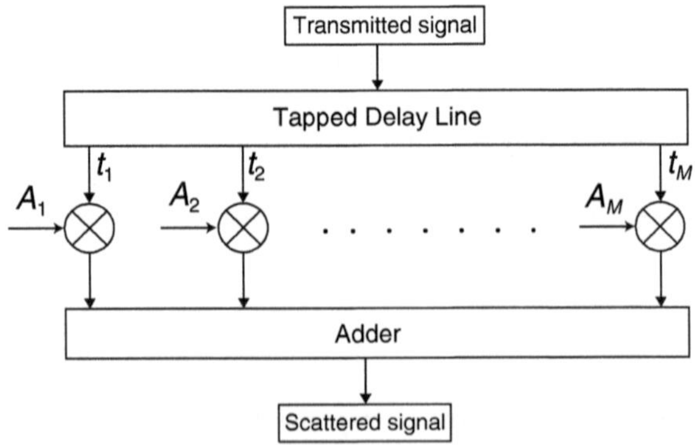

$$e_{\mathrm{e}}(r;t) = \sum_{m=1}^{M} A_{\mathrm{m}} p(t - t_{\mathrm{m}}) \tag{2.50}$$

where A_{m} and t_{m} are the amplitude and time delay related to the mth scattering center. This early-time representation is modeled in Fig. 2.6 as a tapped delay line (t_1, t_2, …, t_{M}) with M multipliers and an adder. This model is compatible with the wavefront representation exhibited in Sect. 2.2.2.3. It was shown that the early-time response emanates from the first interaction of the incident pulse with the scattering centers of the tag, as (2.50) describes.

Based on physical optics approximations, some functions other than impulses must be added to the series in (2.50) in order to completely model the early-time response of the scatterer [40]. The model seen in Fig. 2.7, which includes the parallel combinations of the integrators and differentiators, can be described with the well-accepted model in Fig. 2.6 by assuming that some of the delay differences $t_{\mathrm{m+1}} - t_{\mathrm{m}}$ are very small compared to the smallest wavelength of the impinging signal. If two neighboring scattering centers in Fig. 2.6 have opposite signs, and their delay difference ($d = t_{\mathrm{m+1}} - t_{\mathrm{m}}$) is very small, $d \ll 1$, one can write

$$A_{\mathrm{m}}(p(t - t_{\mathrm{m+1}}) - p(t - t_{\mathrm{m}})) \propto \left.\frac{dp(t)}{dt}\right|_{t=t_{\mathrm{m}}} \tag{2.51}$$

This component is proportional to the differentiated signal at $t = t_m$. Similarly, weighting factors can result in a return component as

$$[p(t) + p(t + d) + \cdots + p(t + kd)] \propto \int_0^{kd} p(t)dt \tag{2.52}$$

In order to perfectly model the early-time response of the scatterer, one must consider both the integrators and differentiators in the model, in addition to the replica of the incident pulse. This model is formulated as the convolution of the incident pulse with the impulse responses of the scattering centers.

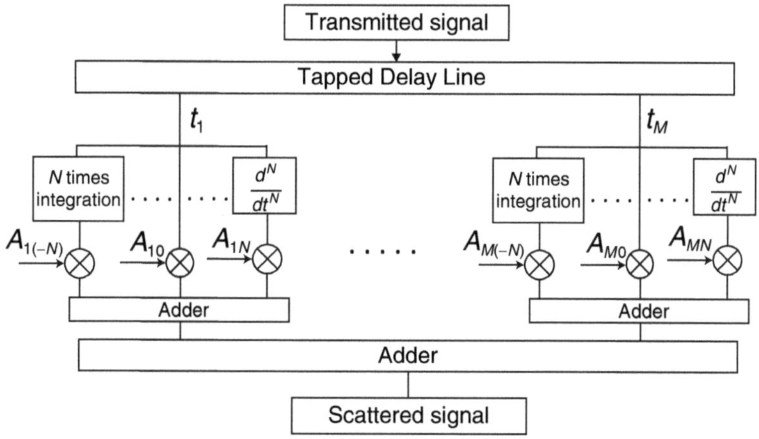

Fig. 2.7 Modified Transversal filter model of the early-time response [40]

$$e_{\mathrm{e}}(r;t) = p(t) * \sum_{m=1}^{M} \sum_{n=-\infty}^{+\infty} A_{\mathrm{mn}} \delta^{(n)}(r; t - t_{\mathrm{m}}) \tag{2.53}$$

In (2.53), the impulse response of the mth scattering center is summed over the integrals and derivatives of the Dirac-delta function. Here, the negative and positive values of n refer to the nth integral and derivative of the delta function with respect to time. In the Laplace domain, the early-time response is written by

$$\begin{aligned} E_{\mathrm{e}}(r;s) &= \sum_{m=1}^{M} \sum_{n=-\infty}^{+\infty} A_{\mathrm{mn}} s^n P(r;s) e^{st_{\mathrm{m}}} \\ &= \sum_{m=1}^{M} \sum_{n=-\infty}^{+\infty} B_{\mathrm{mn}}(r;s) e^{st_{\mathrm{m}}} \end{aligned} \tag{2.54}$$

where

$$B_{\mathrm{mn}} = A_{\mathrm{mn}} s^n P(r;s) \tag{2.55}$$

By comparing (2.54) with (2.46), it is inferred that there is a duality between early-time response in the Laplace domain and late-time response in the time domain. In the former, the response is expanded over the exponential functions of delay times, while in the latter, the time-domain response is expanded versus exponential functions of complex resonances of the tags. This duality will be helpful in the identification process of chipless RFID tags, presented in Chap. 5.

In simple scatterers such as chipless RFID tags, the reflection from the first illuminated part of the tag is strong enough to be considered as the early-time response of the tag. But in the complex scatterers such as an airplane, the impulse responses of the multiple scattering centers of the target should be considered in

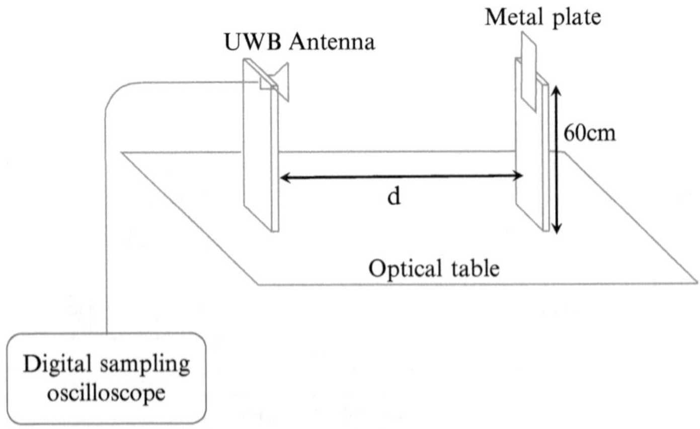

Fig. 2.8 Measurement setup to measure a UWB pulse scattered from a metal object

the response. Therefore, the early-time response from the simple scatterers can accurately be approximated by just one term of the series shown in (2.53). This part is strongly dependent on the polarization and direction of the incident electric field. This is because of the dependency of the scattering centers on the polarization and direction of the incident wave. Additionally, the shape of the early-time response changes from the near field of the scatterer to its far field.

In the near field, the scattered field is mostly similar to the incident pulse, but in the far field, the scattered field is limited to the first time-derivative of the incident field [41].

As an example, the scattered field from a rectangular metal plane with size of 15 cm × 15 cm illuminated by an incident pulse is considered. The measurement set-up is shown in Fig. 2.8. A rectangular metal plane is located 60 cm above an optical table. A TEM horn antenna is connected to a digital sampling oscilloscope in order to calculate the backscattered signal from the tag at different distances. Another measurement without the presence of the metal is performed and the results are subtracted from the earlier signal to cancel the effects of background objects. Time averaging is applied to the received pulses in order to improve the signal-to-noise ratio (SNR). The excitation pulse and its derivative with respect to time are shown in Fig. 2.9. In Fig. 2.10, the backscattered signal from the plate is shown when it is located at four different distances $d = 20$ cm, 30 cm, 1 m and 1.3 m away from the antenna aperture. In the cases where $d = 20$ cm and 30 cm, the observation point is in the near-field of the scatterer and as can be seen, the scattered signal is similar to the incident pulse. In these two cases, the scattered signal is followed by a tail, which is related to the impulse response of the antenna. By locating the plate and the antenna in the far-field of each other, the scattered signal inclines to the first derivative of the incident pulse. By increasing the distance between the antenna and scatterer, the amplitude of the scattered signal decreases. In Fig. 2.11, the normalized responses are plotted for $d = 20$ cm and $d = 130$ cm. According to the results, the scattered field is similar to the incident field in the

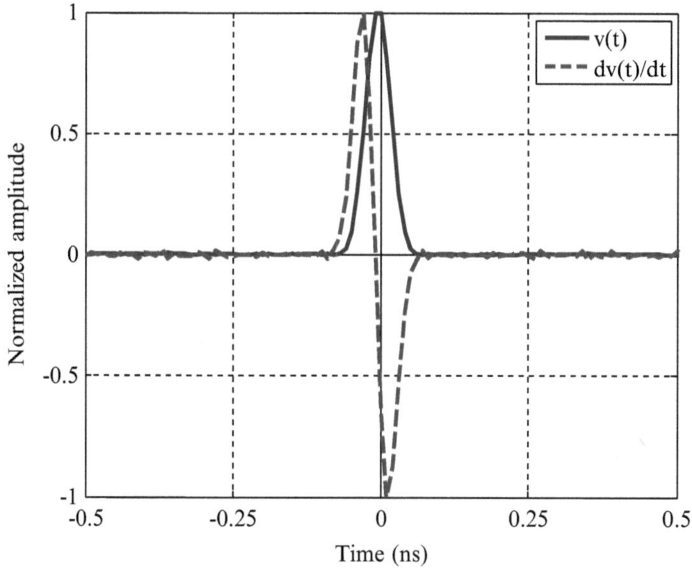

Fig. 2.9 Excitation pulse and its derivative with respect to time

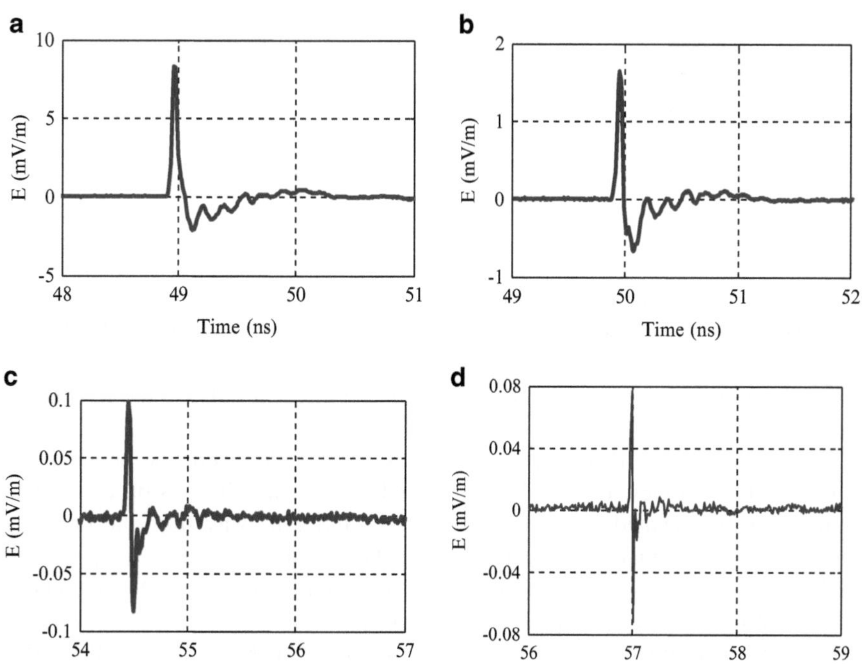

Fig. 2.10 Received electric field from the metal object for different distances (**a**) d = 20 cm, (**b**) d = 30 cm, (**c**) d = 100 cm, and (**d**) d = 130 cm

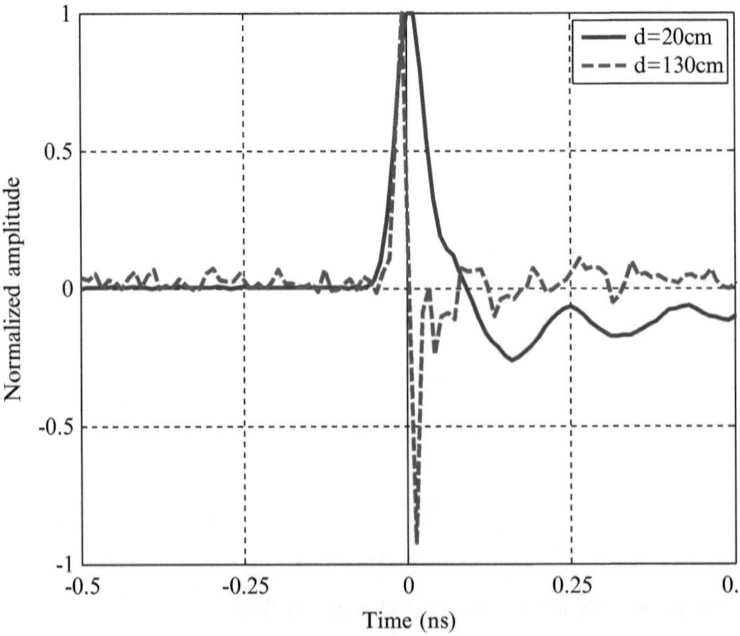

Fig. 2.11 Received electric field from the metal plate for d = 20 cm and d = 130 cm

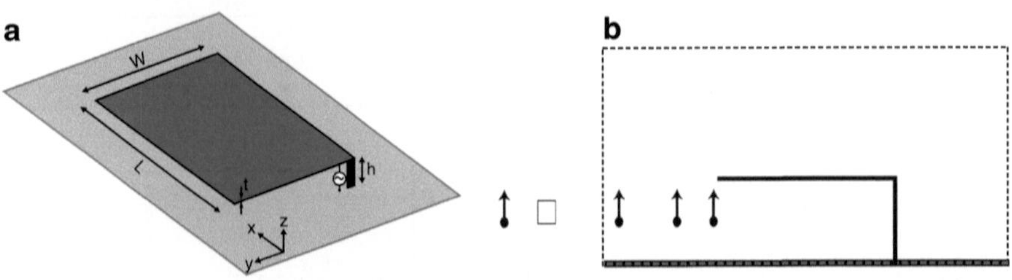

Fig. 2.12 (**a**) 3D view of PIFA antenna (L = 9 mm, W = 5 mm, h = 0.5 mm). (**b**) Cross-section of the antenna and probe located at different distances of the antenna [29] with permission from IEEE

near-field and similar to the first derivative of the incident pulse in the far-field of the scatterer.

This phenomenon also happens in the radiation fields from the antennas. In Fig. 2.12, the 3D view of the planar inverted-F antenna (PIFA) is observed with multiple probes located 1 mm, 10 mm, 30 mm, and 2 m away from the radiating aperture of the antenna. The input signal is a Gaussian pulse with $\tau_{FWHM} = 100ps$ [29]. Figure 2.13 shows the z-component of the radiated fields from the antenna at different probes. As it shows, the early-time response is followed by some sinusoidal signals corresponding to the CNRs of the antenna. Regardless of the probe position at which fields are computed, there are some higher-order resonances at the beginning of all time-domain responses. Each high-order CNR has a unique

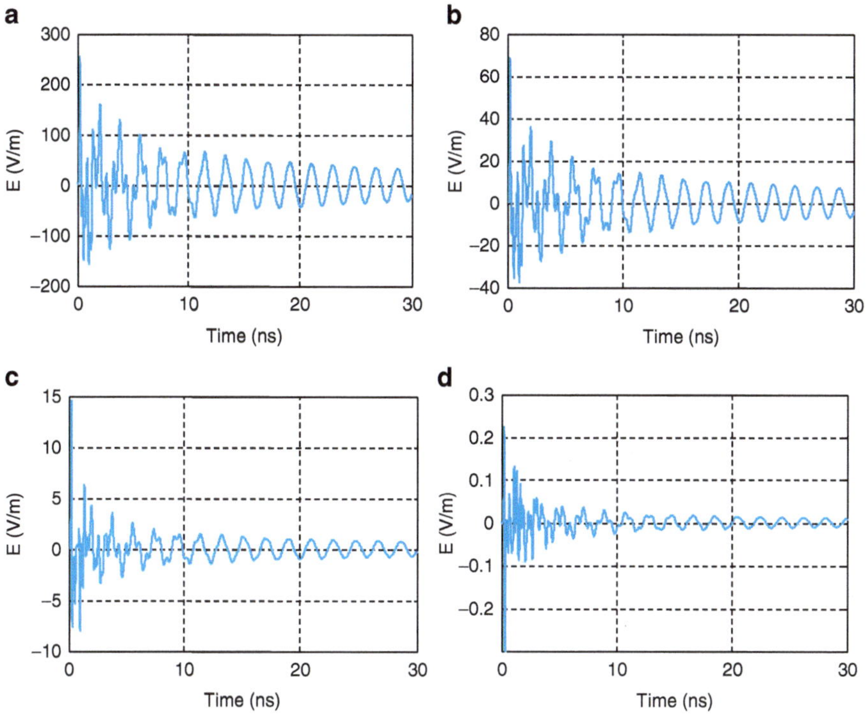

Fig. 2.13 Electric fields received through the probes located at different positions. (**a**) 1 mm (**b**) 10 mm (**c**) 30 mm and (**d**) 2 m [29] with permission from IEEE

damping factor, which is much larger than that of the fundamental mode in this example. Therefore, the effect of the fundamental natural mode is dominant in later times. In Fig. 2.14, the normalized early-time response of the antenna at the probes is depicted. As it shows, at the first probe, located close to the antenna aperture, the electric field is similar to the input pulse. Then by moving farther from the antenna aperture, the shape of the electric field inclines to the first derivative of the input pulse with respect to time [29].

2.2.3 SEM-Based Equivalent Circuit of the Scatterer

In Sect. 2.2.2, the mathematical representation of the scattering process was studied. The early-time response of the scatterer was formulated in (2.53) by the time convolution of the incident pulse with the summation over localized impulse responses of the scattering centers of the tag.

Based on the wavefront method proposed in Sect. 2.2.2.3, the interactions between the local resonances in the early time generate global resonances. According to the singularity expansion method, these global resonances are modeled in the time domain as the summation over damped sinusoidals

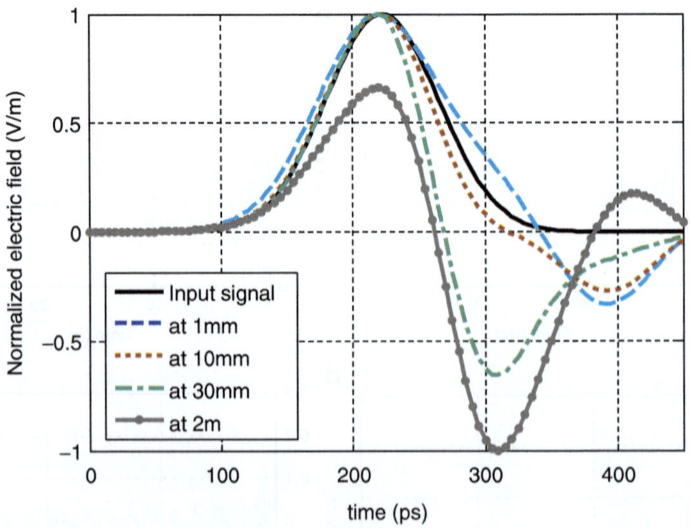

Fig. 2.14 The early-time response of the scatterer received at different probes [29] with permission from IEEE

corresponding to the complex natural resonances of the scatterer with some weighting residues as the coupling coefficients. Although the CNRs are aspect-independent, depending only on the geometry and material of the scatterer, the coupling coefficients are strongly aspect-dependent.

In the scattering analysis, it is usually more desirable to model the scattering process with an equivalent-circuit representation. This equivalent circuit is helpful in the design process of the scatterer. In order to consider the effects of polarization in the scattering process, a general situation, shown in Fig. 2.15, is assumed. Assuming the incident and scattered electric fields as E^{inc} and E^{s} with polarization vectors $\hat{a}_{inc}$ and $\hat{a}_s$, the scattering transfer function of the tag is defined as

$$H_{\mathrm{t}}(\hat{a}_{\mathrm{inc}}, \hat{a}_{\mathrm{s}}; s) = \frac{E^{\mathrm{s}}(r; s)}{E^{\mathrm{inc}}(r; s)} \tag{2.56}$$

Assuming the incident electric field as a Dirac-delta function, the transfer function of the tag is related to the scattered electric field. The equivalent circuit of the tag, shown in Fig. 2.15, is depicted in Fig. 2.16. The input and output voltages are defined at the transmitting and receiving antenna ports, respectively. The incident field is coupled to the CNRs by coupling coefficients $n_{\mathrm{inc}}^{(\mathrm{i})}$ $i = 1, 2, \ldots, N$, which depend on the direction and polarization of the incident electric field.

Each CNR is represented by a parallel RLC circuit in series with a delay line, which models the turn-on times of the CNRs. The excited natural current modes are coupled to the scattered field with coupling coefficients $n_{\mathrm{s}}^{(\mathrm{i})}$ $i = 1, 2, \ldots, N$. The quantity Z_e represents the early-time response of the tag, which is aspect-dependent. The transfer function of the tag as defined in (2.56) is converted to the ratio of

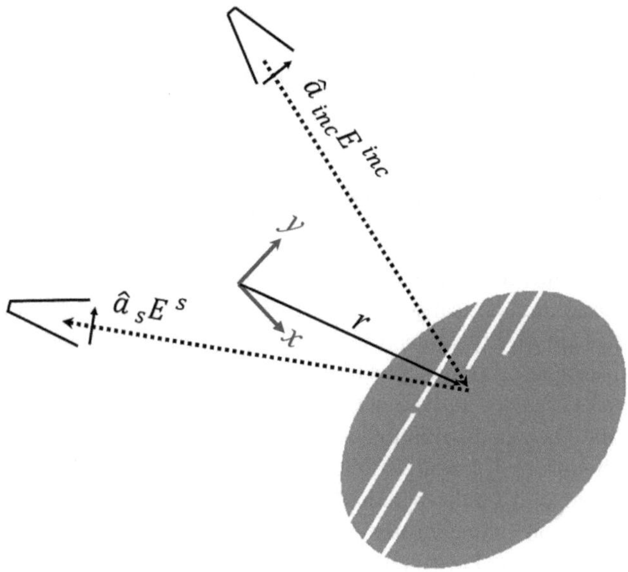

Fig. 2.15 A tag illuminated by an incident wave

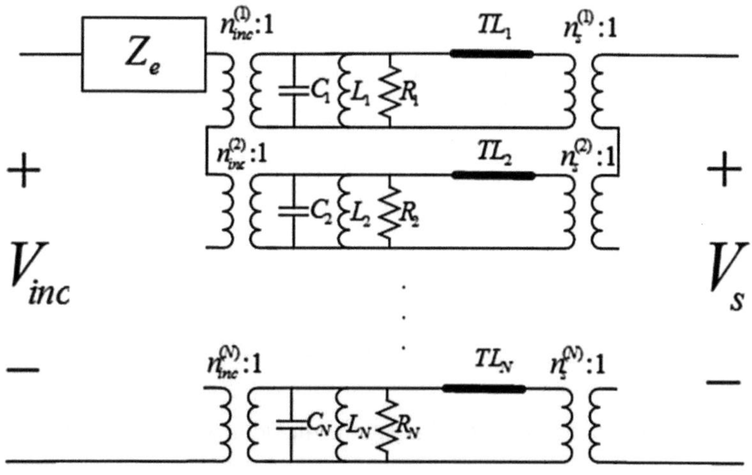

Fig. 2.16 SEM-based equivalent circuit of chipless RFID tag

output to input voltages. With some mathematical manipulation, the transfer function can easily be written as

$$\begin{aligned} H(s) &= H_{\mathrm{e}}(s) + H_{\mathrm{l}}(s) \\ &= H_{\mathrm{e}}(s) + \sum_{n=1}^{N}\left[\frac{A_{\mathrm{n}}}{s - s_n} + \frac{A_n{}^*}{s - s_n{}^*}\right] \end{aligned} \tag{2.57}$$

The first term is the early-time part of the tag and the second term, including the complex natural resonances, is the late-time part. Although in reality, N is infinity, for numerical computations, N is usually truncated to a finite value.

2.2.4 Example: Scattering from a Dipole

In this section, scattered fields from a dipole are formulated based on the SEM in order to illustrate the application of the SEM in numerical computations [42]. A single dipole of length L is aligned with the z-axis and illuminated by an incident plane wave propagating in the direction forming an angle θ with the z-axis. The incident wave is assumed to be a step function, striking the scatterer at $t=0$. By formulating the current distribution using the SEM, it is possible to obtain the current distribution and scattered fields in the time domain. By neglecting the effects of the end caps on the wire and φ variations of the currents on the wire, a Pocklington equation can be written for the axially-directed current on the dipole. Assuming $s=\alpha+j\omega$, the Pocklington equation is written as [42]

$$-s\varepsilon_0 E_t^{\mathrm{inc}}(z,s) = \int_{-L/2}^{L/2} I\left(z',s\right)\left(\frac{\partial^2}{\partial z^2}-\frac{s^2}{c^2}\right)K\left(z,z';s\right)dz' \tag{2.58}$$

where E_t^{inc} is the tangential component of the incident electric field along the dipole and the kernel K is given by

$$K\left(z,z';s\right) = \frac{1}{2\pi a}\int_0^{2\pi}\frac{\exp(-sR/c)}{4\pi R}ad\phi \tag{2.59}$$

Here, a is the radius of the wire and

$$R = \left[\left(z-z'\right)^2 + 4a^2\sin^2(\varphi/2)\right]^{1/2} \tag{2.60}$$

The incident tangential electric field along the dipole is written by (Fig. 2.17)

$$E_t^{inc}(z,s) = E^{inc}(s)\cos(\theta)\exp\left(\frac{sz\sin(\theta)}{c}\right) \tag{2.61}$$

For the step-function incident wave, we have

$$E^{inc}(s) = \frac{E_0}{s} \tag{2.62}$$

By discretizing the length of the dipole into N segments, the integral equation in (2.58) is converted to the equation

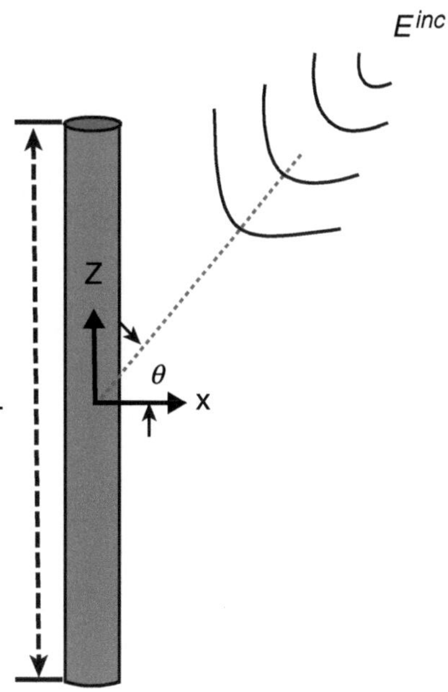

Fig. 2.17 Geometry of the dipole illuminated by an incident field [42] with permission from IEEE

$$[Z][I] = [V] \tag{2.63}$$

where $[Z]$ is an $N \times N$ matrix referred to as the system impedance matrix, $[I]$ is an $N \times 1$ response vector and $[V]$ is an $N \times 1$ vector corresponding to the incident field. According to (2.30), at CNRs of the scatterer, the following equation holds

$$[Z(s_n)][I(s_n)] = 0 \tag{2.64}$$

The CNRs of the dipole are obtained from

$$\Delta(s_n) = \det(Z(s_n)) = 0 \tag{2.65}$$

The CNRs can be calculated by employing different searching algorithms. One easy way is to expand $\Delta(s)$ in a complex Taylor series about s_n as [42]

$$\Delta(s_n) = \Delta(s_0) + \Delta'(s_0)(s - s_0) + \cdots = 0 \tag{2.66}$$

Keeping the first two terms of the series, the CNR, s_n, is obtained from

$$s_n = s_0 - \frac{\Delta(s_0)}{\Delta'(s_0)} \tag{2.67}$$

where s_0 is the initial guess of the resonant frequency. More accurate values for s_n can be obtained by repeating this procedure. Figure 2.18 shows the pole diagram of

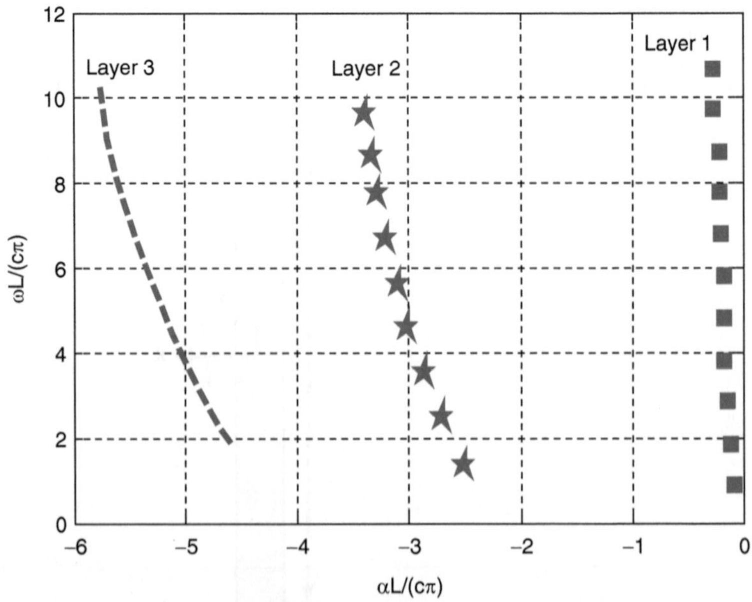

Fig. 2.18 Pole diagram of the dipole, representing the resonant frequency and damping factor of the CNRs [42] with permission from IEEE

the dipole normalized to $c\pi/L$. It is seen that the poles, $s_n = \alpha_n + j\omega_n$, are located in different layers in the s-plane. The poles situated in the first layer are more dominant in the time-domain response because they have lower damping factors than those located in the further layers. The natural current modes on the dipole are the solutions of equation (2.63).

$$[I] = [Z]^{-1}[V] = \frac{[Y]}{\Delta(s)}[V] \tag{2.68}$$

In Fig. 2.19, the real and imaginary parts of the first three modes of the dipole located in the first layer are depicted.

By possessing the natural modes and CNRs of the dipole, the current distribution can be cast to the form

$$[I] = \sum_n \frac{R_n}{s - s_n}[J_n] \tag{2.69}$$

where R_n is the residue of the nth pole, obtained from (2.34). The time-domain response is obtained by applying an inverse Laplace transform along the appropriate Bromwich contour.

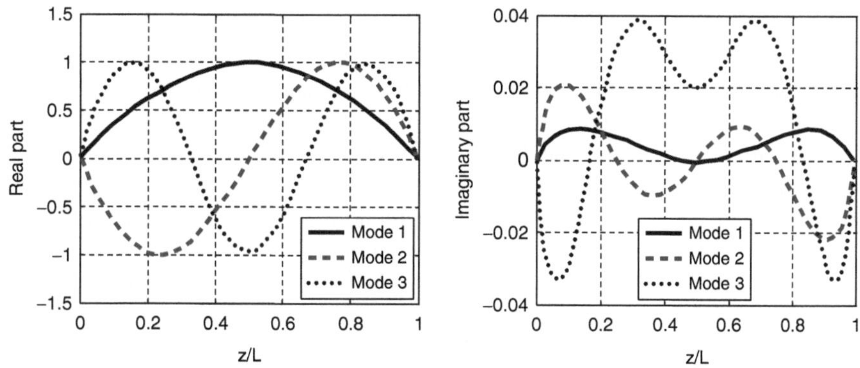

Fig. 2.19 Real and imaginary parts of the first three natural currents of the dipole [42] with permission from IEEE

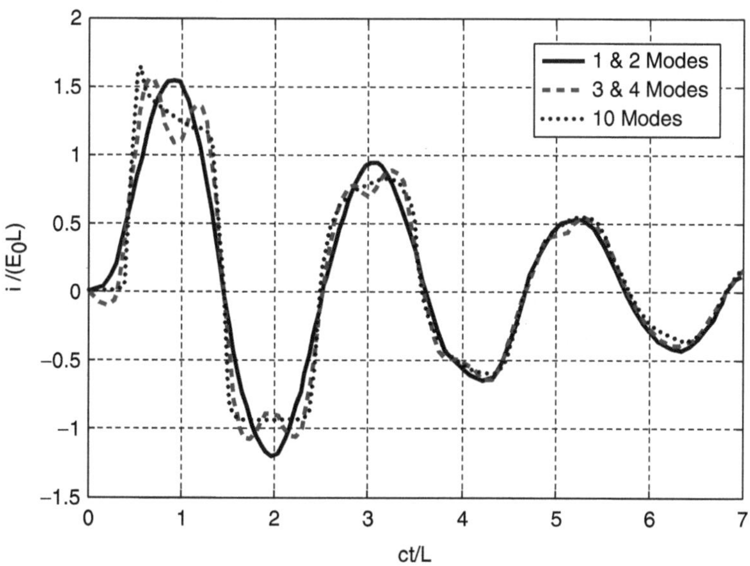

Fig. 2.20 Current distribution at z/L = 0.5 as a function of time [42] with permission from IEEE

$$[i(t)] = \sum_{n} R_n \exp(s_n t)[J_n] \tag{2.70}$$

We consider the case where the incident electric field propagates in the direction forming $\theta = 30°$ with the dipole. The current at $z/L = 0.5$ is shown in Fig. 2.20 for different numbers of modes considered in the summation. Here, just the poles in the first layer are considered. The effect of high-order resonances is clearly seen in the time-domain response of the dipole. Since in this case, by increasing the order of the resonant mode, its damping factor increases, we can infer that its effect is dominant in earlier times of the response.

2.3 Eigenmode Expansion Method

In the previous section, the scattered fields from the tag were expanded versus the singularities of the structure. Another way of representing the scattered fields from the scatterer is to expand the fields or currents versus the eigenmodes of the structure under consideration. The question may arise of how to extract the singularity poles of the scatterer from the eigenmode expansion of the fields. Before starting the general formulation of the scattered fields from an arbitrary scatterer, a simple case shown in Fig. 2.21 is considered. An ideal dipole is located in a hollow rectangular cavity resonator. In the case shown in Fig. 2.21, the eigenmodes of the structure are easily written in a closed form. By calculating the dyadic Green's function of the structure, the scattered fields are obtained from (2.14). The electric and magnetic fields in the cavity satisfy the following equations

$$\nabla \times \nabla \times \mathbf{E} - k^2\mathbf{E} = -j\omega\mu\mathbf{J} \tag{2.71}$$

$$\nabla \times \nabla \times \mathbf{H} - k^2\mathbf{H} = \nabla \times \mathbf{J} \tag{2.72}$$

with the following boundary conditions on the walls.

$$\hat{n} \times \mathbf{E} = 0 \tag{2.73}$$

$$\hat{n} \cdot \mathbf{H} = 0 \tag{2.74}$$

Quantity $\hat{n}$ indicates the direction normal to the wall surfaces. In order to solve the preceding boundary-value problem, two quantities, the electric-type dyadic

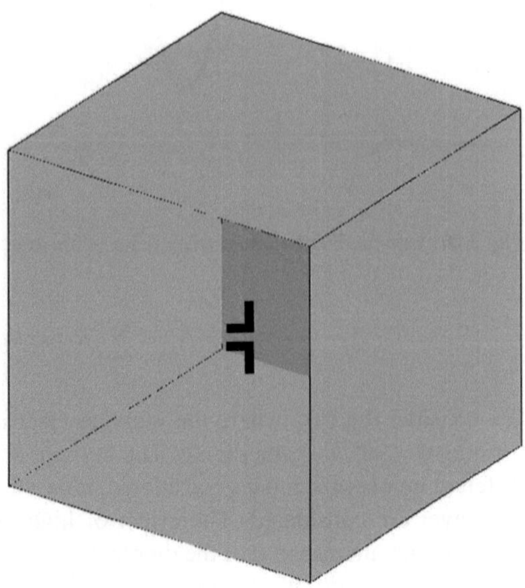

Fig. 2.21 An ideal dipole in a rectangular cavity resonator

Green's function, $\overleftrightarrow{\mathbf{G}}_\mathrm{e}$, and the magnetic-type dyadic Green's function, $\overleftrightarrow{\mathbf{G}}_\mathrm{m}$, are defined as [43]

$$\nabla\times\nabla\times\overleftrightarrow{\mathbf{G}}_\mathrm{e}-k^2\overleftrightarrow{\mathbf{G}}_\mathrm{e}=\overleftrightarrow{\mathbf{I}}\delta\left(\mathbf{r}-\mathbf{r}'\right) \tag{2.75}$$

$$\nabla\times\nabla\times\overleftrightarrow{\mathbf{G}}_\mathrm{m}-k^2\overleftrightarrow{\mathbf{G}}_\mathrm{m}=\nabla\times\overleftrightarrow{\mathbf{I}}\delta\left(\mathbf{r}-\mathbf{r}'\right) \tag{2.76}$$

with the following boundary conditions on the walls

$$\hat{n}\times\overleftrightarrow{\mathbf{G}}_\mathrm{e}=0 \tag{2.77}$$

$$\hat{n}\cdot\overleftrightarrow{\mathbf{G}}_\mathrm{m}=0 \tag{2.78}$$

The quantity $\overleftrightarrow{\mathbf{I}}$ in (2.75) and (2.76) is the unit dyad. It is interesting to note that the following relation holds between $\overleftrightarrow{\mathbf{G}}_\mathrm{e}$ and $\overleftrightarrow{\mathbf{G}}_\mathrm{m}$ [44]

$$k^2\overleftrightarrow{\mathbf{G}}_\mathrm{e}=\nabla\times\overleftrightarrow{\mathbf{G}}_\mathrm{m}-\overleftrightarrow{\mathbf{I}}\delta\left(\mathbf{r}-\mathbf{r}'\right) \tag{2.79}$$

From (2.75) and (2.76), it is clear that $\nabla\cdot\overleftrightarrow{\mathbf{G}}_\mathrm{e}\neq 0$ and $\nabla\cdot\overleftrightarrow{\mathbf{G}}_\mathrm{m}=0$, while the cavity modes are divergenceless. Therefore, the magnetic dyadic Green's function can be expanded versus cavity modes, and by using the relation in (2.79), the electric dyadic Green's function is obtained. From $\nabla\cdot\overleftrightarrow{\mathbf{G}}_\mathbf{m}=0$ and $\nabla\times\nabla\times=\nabla\nabla\cdot-\nabla^2$, the following alternative form of (2.76) is obtained.

$$\left(\nabla^2+k^2\right)\overleftrightarrow{\mathbf{G}}_\mathrm{m}=-\nabla\times\overleftrightarrow{\mathbf{I}}\delta\left(\mathbf{r}-\mathbf{r}'\right) \tag{2.80}$$

$$\left.\begin{aligned}\hat{n}\cdot\overleftrightarrow{\mathbf{G}}_\mathrm{m}&=0\\ \hat{n}\times\nabla\times\overleftrightarrow{\mathbf{G}}_\mathrm{m}&=0\end{aligned}\right\}\text{on the wall} \tag{2.81}$$

By defining another Green's function as

$$\left(\nabla^2+k^2\right)g_\mathrm{m}=-\mathbf{I}\delta\left(\mathbf{r}-\mathbf{r}'\right) \tag{2.82}$$

$$\left.\begin{aligned}\hat{n}\cdot g_\mathrm{m}&=0\\ \hat{n}\times\nabla\times g_\mathrm{m}&=0\end{aligned}\right\}\text{on the walls} \tag{2.83}$$

and applying Green's theorem, we have

$$\overleftrightarrow{G}_{\mathrm{m}}\left(r,r'\right) = \int g_{\mathrm{m}}\left(r,r''\right)\cdot\nabla''\times\mathbf{I}\delta\left(r'-r''\right)ds'' \tag{2.84}$$

Where ∇'' is differentiation versus r''. Therefore, the components of the dyadic function in (2.82) can be written in the following form

$$(\nabla^2+k^2)\begin{Bmatrix} g_{\mathrm{m}}^{xx} \\ g_{\mathrm{m}}^{yy} \\ g_{\mathrm{m}}^{zz} \end{Bmatrix} = -\begin{Bmatrix} 1 \\ 1 \\ 1 \end{Bmatrix}\delta\left(\mathbf{r}-\mathbf{r}'\right) \tag{2.85}$$

By expanding the dyadic components versus cavity modes and applying boundary conditions, one can obtain

$$g_{\mathrm{m}}\left(r,r'\right) = \sum_{n=0}^{\infty}\sum_{m=0}^{\infty}\sum_{l=0}^{\infty}\frac{\varepsilon_{0n}\varepsilon_{0m}\varepsilon_{0l}}{abc\left[k^2-\left(\frac{n\pi}{a}\right)^2-\left(\frac{m\pi}{b}\right)^2-\left(\frac{l\pi}{c}\right)^2\right]}\cdot$$
$$\left\{\hat{x}\hat{x}\left(\sin\left(\frac{n\pi x}{a}\right)\sin\left(\frac{n\pi x'}{a}\right)\cos\left(\frac{m\pi y}{b}\right)\cos\left(\frac{m\pi y'}{b}\right)\cos\left(\frac{l\pi z}{c}\right)\cos\left(\frac{l\pi z'}{c}\right)\right)\right.$$
$$+\hat{y}\hat{y}\left(\cos\left(\frac{n\pi x}{a}\right)\cos\left(\frac{n\pi x'}{a}\right)\sin\left(\frac{m\pi y}{b}\right)\sin\left(\frac{m\pi y'}{b}\right)\cos\left(\frac{l\pi z}{c}\right)\cos\left(\frac{l\pi z'}{c}\right)\right)$$
$$\left.+\hat{z}\hat{z}\left(\cos\left(\frac{n\pi x}{a}\right)\cos\left(\frac{n\pi x'}{a}\right)\cos\left(\frac{m\pi y}{b}\right)\cos\left(\frac{m\pi y'}{b}\right)\sin\left(\frac{l\pi z}{c}\right)\sin\left(\frac{l\pi z'}{c}\right)\right)\right\} \tag{2.86}$$

where $k^2=\omega^2\mu\varepsilon$. Using (2.84) and relation (2.79), the total dyadic Green's functions of the cavity can be obtained. Assuming the current element is oriented in the y-direction, the electric-type dyadic component is obtained as

$$\mathbf{G}_{\mathrm{e}}^{yy}\left(\mathbf{r},\mathbf{r}'\right) = \frac{1}{k^2}\sum_{n=0}^{\infty}\sum_{m=0}^{\infty}\sum_{l=0}^{\infty}\frac{\varepsilon_{0n}\varepsilon_{0m}\varepsilon_{0l}\left[\left(\frac{l\pi}{c}\right)^2+\left(\frac{n\pi}{a}\right)^2\right]}{abc\left[k^2-\left(\frac{n\pi}{a}\right)^2-\left(\frac{m\pi}{b}\right)^2-\left(\frac{l\pi}{c}\right)^2\right]}\cdot$$
$$\left(\sin\left(\frac{n\pi x}{a}\right)\sin\left(\frac{n\pi x'}{a}\right)\cos\left(\frac{m\pi y}{b}\right)\cos\left(\frac{m\pi y'}{b}\right)\sin\left(\frac{l\pi z}{c}\right)\sin\left(\frac{l\pi z'}{c}\right)\right)-\frac{1}{k^2}\delta\left(r-r'\right) \tag{2.87}$$

In (2.87), the yy-component of the electric-type dyadic Green's function and consequently field distributions in the cavity are expanded versus the cavity modes in a simple closed-form relation. The cavity modes satisfy the following eigenvalue equation

$$\begin{aligned}\mathcal{L}(E) &= \left(\frac{\partial^2}{\partial x^2}+\frac{\partial^2}{\partial y^2}+\frac{\partial^2}{\partial z^2}+\omega^2\mu\varepsilon\right)E(x,y,z) \\ &= \lambda(\omega)\cdot E(x,y,z)\end{aligned} \tag{2.88}$$

where

$$\lambda(\omega) = \left(\omega^2\mu\varepsilon - \left(\frac{n\pi}{a}\right)^2 - \left(\frac{m\pi}{b}\right)^2 - \left(\frac{l\pi}{c}\right)^2\right) \tag{2.89}$$

represents the eigenvalue of the equation (2.88) and $m, n, l = 0, 1, 2, \ldots$. By sorting the eigenvalues with index n and representing the eigenmodes of the operator $\mathcal{L}$ in (2.88) by $\mathbf{J}_n(\mathbf{r};s)$, the Green's function in (2.88) in source-free regions can be written by

$$\overleftrightarrow{\mathbf{G}}_e\left(\mathbf{r},\mathbf{r}';s\right) = \sum_n \frac{1}{\lambda_n(s)} A_n \mathbf{J}_n(\mathbf{r};s)\mathbf{J}_n\left(\mathbf{r}';s\right) \tag{2.90}$$

According to (2.89) and (2.90), the singularity poles of the structure, which are the resonant frequencies of the cavity, are the zeroes of the eigenvalues in (2.89). It will be shown that the scattered field presented in (2.90) can be generalized for any arbitrary scattering scenario.

Returning to scattering from the chipless tag shown in Fig. 2.2, the incident electric field induces a current distribution on the tag, which can be calculated by applying boundary conditions on the tag surface as

$$\left\langle \overleftrightarrow{\mathbf{G}}\left(\mathbf{r},\mathbf{r}';s\right), \mathbf{J}\left(\mathbf{r}'\right)\right\rangle_{\mathbf{r}'} = -\mathbf{E}_t^{inc}(\mathbf{r};s) \qquad \forall \mathbf{r}\in A \tag{2.91}$$

Where $\hat{\mathbf{G}}$ in the following formulations is the electric-type dyadic Green's function and A represents the surface of the tag. The eigenvalue equation associated with (2.91) is written by

$$\left\langle \overleftrightarrow{\mathbf{G}}\left(\mathbf{r},\mathbf{r}';s\right), \mathbf{J}_n\left(\mathbf{r}';s\right)\right\rangle_{\mathbf{r}'} = \lambda_n(s)\mathbf{J}_n(\mathbf{r};s) \qquad \forall \mathbf{r}\in A \tag{2.92}$$

where $\mathbf{J}_n(\mathbf{r};\, s)$ and $\lambda_n(s)$ are nth eigenmode and eigenvalue of $\overleftrightarrow{\mathbf{G}}$, respectively. By applying the method of moment (MoM), the integral equation in (2.91) is converted to the following matrix equation

$$\boldsymbol{\Gamma}(s)\cdot\mathbf{J}(s) = \mathbf{I}_e(s) \tag{2.93}$$

and the eigenvalue equation corresponding to (2.93) is written by

$$\boldsymbol{\Gamma}(s) \cdot \mathbf{J}_{\mathrm{n}}(s) = \lambda_{\mathrm{n}}(s)\mathbf{J}_{\mathrm{n}}(s) \tag{2.94}$$

In order to have nontrivial solutions, the determinant of the coefficient matrix must be zero as

$$C = \det(\boldsymbol{\Gamma}(s) - \lambda_{\mathrm{n}}(s)\mathbf{I}) = 0 \tag{2.95}$$

C, is called the characteristic equation and $\mathbf{I}$, is a unit matrix. Assuming $\boldsymbol{\Gamma}$ to be a square matrix of rank N, one can write

$$\det(\boldsymbol{\Gamma}(s)) = \prod_{n=1}^{N} \lambda_{\mathrm{n}}(s) \tag{2.96}$$

The eigenvalues may or may not all be distinct. It is clearly seen from (2.95) and (2.96) that the CNRs of the scatterer are the zeroes of the eigenvalues. Each eigenvalue may contain an infinite number of CNRs. For each square matrix, two sets of eigenmodes, right-side and left-side, are defined. In (2.94), the right-side eigenmodes are introduced, which are represented by $\mathbf{J}_{\mathrm{n}}^{\mathrm{R}}$ in the following. The left-side Eigenmodes are defined as [45]

$$\mathbf{J}_{\mathrm{n}}^{\mathrm{L}}(s) \cdot \boldsymbol{\Gamma}(s) = \lambda_{\mathrm{n}}(s)\mathbf{J}_{\mathrm{n}}^{\mathrm{L}}(s) \tag{2.97}$$

The orthogonality and bi-orthogonality relations between eigenmodes can be summarized as [45]

$$\mathbf{J}_{\mathrm{n}}^{\mathrm{L}}(s) \cdot \mathbf{J}_{\mathrm{m}}^{\mathrm{L}}(s) = \delta_{\mathrm{mn}} \tag{2.98a}$$

$$\mathbf{J}_{\mathrm{n}}^{\mathrm{R}}(s) \cdot \mathbf{J}_{\mathrm{m}}^{\mathrm{R}}(s) = \delta_{\mathrm{mn}} \tag{2.98b}$$

$$\mathbf{J}_{\mathrm{n}}^{\mathrm{L}}(s) \cdot \mathbf{J}_{\mathrm{m}}^{\mathrm{R}}(s) = \delta_{\mathrm{mn}} \tag{2.98c}$$

where

$$\delta_{\mathrm{mn}} = \begin{cases} 0 & m \neq n \\ 1 & m = n \end{cases}$$

The current distribution and incident electric field in (2.94) can be expanded versus the eigenmodes as

$$\mathbf{J}(s) = \sum_{n=1} a_{\mathrm{n}}\mathbf{J}_{\mathrm{n}}^{\mathrm{R}}(s) \tag{2.99}$$

$$\mathbf{I}_{\mathrm{e}} = \sum_{n=1} b_{\mathrm{n}}\mathbf{J}_{\mathrm{n}}^{\mathrm{R}}(s) \tag{2.100}$$

Substituting (2.99) into (2.93) and using (2.97), one arrives at

$$\sum_{n=1} a_{\mathrm{n}}\boldsymbol{\Gamma}(s)\cdot\mathbf{J}_{\mathrm{n}}^{(\mathrm{R})}(s)=\sum_{n=1} a_{\mathrm{n}}\lambda_{\mathrm{n}}(s)\mathbf{J}_{\mathrm{n}}^{(\mathrm{R})}(s)\\ =\mathbf{I}_{\mathrm{e}}(s) \tag{2.101}$$

By taking an inner product of the two sides of (2.101) with $\mathbf{J}_{\mathrm{n}}^{(\mathrm{L})}(s)$ and using the orthogonality relation in (2.98a, 2.98b and 2.98c), one can write

$$a_{\mathrm{n}}=\frac{1}{\lambda_{\mathrm{n}}(s)}\cdot\frac{\mathbf{J}_{\mathrm{n}}^{(\mathrm{L})}(s)\cdot\mathbf{I}(s)}{\mathbf{J}_{\mathrm{n}}^{(\mathrm{L})}(s)\cdot\mathbf{J}_{\mathrm{n}}^{(\mathrm{R})}(s)} \tag{2.102}$$

Therefore, the current distribution on the tag is written by

$$\mathbf{J}(s)=\sum_{n=1}\frac{1}{\lambda_{\mathrm{n}}(s)}\cdot\frac{\mathbf{J}_{\mathrm{n}}^{(\mathrm{L})}(s)\cdot\mathbf{I}(s)}{\mathbf{J}_{\mathrm{n}}^{(\mathrm{L})}(s)\cdot\mathbf{J}_{\mathrm{n}}^{(\mathrm{R})}(s)}\mathbf{J}_{\mathrm{n}}^{(\mathrm{R})}(s) \tag{2.103}$$

By defining the normalized dyadic functions as

$$\mathbf{d}_{\mathrm{n}}(s)=\frac{\mathbf{J}_{\mathrm{n}}^{(\mathrm{R})}(s)\mathbf{J}_{\mathrm{n}}^{(\mathrm{L})}(s)}{\mathbf{J}_{\mathrm{n}}^{(\mathrm{R})}(s)\cdot\mathbf{J}_{\mathrm{n}}^{(\mathrm{L})}} \tag{2.104}$$

the current distribution in (2.103) can be expressed by

$$\mathbf{J}(s)=\sum_{n}\frac{1}{\lambda_{\mathrm{n}}(s)}\mathbf{d}_{\mathrm{n}}(s)\cdot\mathbf{I}(s) \tag{2.105}$$

By comparing (2.93) and (2.105), one can write

$$\boldsymbol{\Gamma}^{-1}=\sum_{n}\frac{1}{\lambda_{\mathrm{n}}(s)}\mathbf{d}_{\mathrm{n}}(s) \tag{2.106}$$

It shows that the singularity poles of the scatterer are the zeroes of the Eigenvalues. Similarly,

$$\boldsymbol{\Gamma}=\sum_{n}\lambda_{\mathrm{n}}(s)\mathbf{d}_{\mathrm{n}}(s) \tag{2.107}$$

Therefore,

$$(\boldsymbol{\Gamma}(s))^{-1}\cdot\boldsymbol{\Gamma}(s)=\sum_{n}\mathbf{d}_{\mathrm{n}}\\ =\boldsymbol{\delta} \tag{2.108}$$

In some scattering problems, the geometry of the scatterer is perfectly matched to a specific coordinate system. As an example, when the incident electric field illuminates a perfectly electric conducting (PEC) sphere or an infinite cylinder, the scattered fields or equivalently the induced currents on the scatterer can be easily expanded versus the eigenmodes of the structures [45, 46]. For arbitrary geometries, which are not necessarily compatible with any specific coordinate system, the numerical evolution of the eigenmode equation (2.97) can be used in order to find the eigenmodes, eigenvalues and, consequently, the complex natural resonances of the scatterer.

2.4 Theory of Characteristic Modes

In the previous section, the current and consequently, the radiated fields from the scatterer were expanded versus the eigenmodes of the dyadic Green's function of the structure in the complex frequency domain. The resulted matrix from MoM is not Hermitian and its eigenmodes and eigenvalues are complex. For well-coordinated structures such as sphere, the eigenmodes of the dyadic Green's function are in-phase on the surface of the scatterer in the spherical coordinate system. By changing the geometry of the scatterer, the eigen-currents will not be in-phase anymore. The question is how to extend the idea of in-phase basis functions for arbitrary-shape scatterer. The theory of characteristic modes introduces a weighted eigenvalue equation whose eigenvalues are in-phase on the scatterer surface. This theory was first introduced by Garbacz in 1971 for conducting bodies of arbitrary shapes [47]. His proposed approach was based on diagonalizing the scattering matrix of the scatterers. He presented a new class of eigenmodes on a scatterer that are real and their corresponding scattered fields have constant phase over the surface of the body. Although the proposed method was used in some cases, its implementation was not easy for an arbitrarily-shaped scatterer. In [48], Harrington proposed an alternative viewpoint for diagonalizing an operator. This technique relates the current distribution to the tangential electric field on the body. He defined a particular weighted eigenvalue equation, which gives the same eigenmodes defined by Garbacz, but with a simpler approach. Ever since, this proposed technique has been widely employed in the design and modeling of antennas and scatterers [49–52]. Similar to natural resonant modes, the characteristic modes are independent of source fields and depend only on the geometry and shape of a scatterer.

2.4.1 Mathematical Formulation of the Characteristic Mode Theory

Referring to Fig. 2.2, it is assumed that an incident plane wave illuminates the tag. The first step in formulating the eigenvalue equation, which defines the characteristic modes of the tag, is the application of the electric-field integral equation (EFIE) on the tag surface as

$$\begin{aligned}\mathcal{L}(J) &= \left\langle \overleftrightarrow{\mathbf{G}}\left(\mathbf{r},\mathbf{r}';s\right),\mathbf{J}\left(\mathbf{r}'\right)\right\rangle_{\mathbf{r}'} \\ &= -\mathbf{E}_{\mathrm{t}}^{\mathrm{inc}}(\mathbf{r};s)\end{aligned} \tag{2.109}$$

where $\overleftrightarrow{\mathbf{G}}$ is defined in (2.15) and the integration is performed over the surface of the tag. Since the operator $\mathcal{L}(.)$ in (2.109) has the dimensions of impedance, it is more convenient to introduce the notation

$$\mathcal{Z}(\mathbf{J}) = \mathcal{L}(\mathbf{J}) \tag{2.110}$$

where $\mathcal{Z}$ is a symmetric operator as

$$\langle B, \mathcal{Z}(C)\rangle = \langle \mathcal{Z}(B), C\rangle \tag{2.111}$$

One can write $\mathcal{Z}$ in terms of its real and imaginary components

$$\mathcal{Z} = \mathcal{R} + j\mathcal{X} \tag{2.112}$$

where

$$\begin{aligned}\mathcal{R} &= \frac{1}{2}(\mathcal{Z} + \mathcal{Z}^*) \\ \mathcal{X} &= \frac{1}{2j}(\mathcal{Z} - \mathcal{Z}^*)\end{aligned}$$

Since the radiated power from a current distribution $\mathbf{J}$ on the tag is given by

$$P = \langle \mathbf{J}^*, \mathcal{R}(\mathbf{J})\rangle \tag{2.113}$$

it follows that $\mathcal{R}$ is positive semi-definite. The starting step in defining the characteristic modes of the tag is the following eigenvalue equation

$$\mathcal{Z}(\mathbf{J}_{\mathrm{n}}) = \gamma_{\mathrm{n}}\mathcal{M}(\mathbf{J}_{\mathrm{n}}) \tag{2.114}$$

where γ_{n} and $\mathbf{J}_{\mathrm{n}}$ are the nth eigenvalue and eigenmode of the equation and $\mathcal{M}$ is a weighting operator. Choosing $\mathcal{M} = \mathcal{R}$ ensures orthogonality of the radiation patterns of the current modes in the far zone. Introducing

$$\gamma_{\mathrm{n}} = 1 + j\lambda_{\mathrm{n}} \tag{2.115}$$

into (2.114) , the eigenvalue equation is converted to

$$\mathcal{X}(\mathbf{J}_n) = \lambda_n \mathcal{R}(\mathbf{J}_n) \tag{2.116}$$

Since $\mathcal{R}$ and $\mathcal{X}$ are real symmetric operators, both the eigenvalues λ_n and the corresponding characteristic modes, $\mathbf{J}_n$, must be real. In addition, the eigenmodes satisfy the orthogonality relationships

$$\langle \mathbf{J}_m, \mathcal{R}(\mathbf{J}_n) \rangle = \delta_{mn} \tag{2.117a}$$

$$\langle \mathbf{J}_m, \mathcal{X}(\mathbf{J}_n) \rangle = \lambda_n \delta_{mn} \tag{2.117b}$$

$$\langle \mathbf{J}_m, \mathcal{Z}(\mathbf{J}_n) \rangle = (1 + j\lambda_n)\delta_{mn} \tag{2.117c}$$

where

$$\delta_{mn} = \begin{cases} 1 & m = n \\ 0 & m \neq n \end{cases}$$

2.4.2 Characteristic Fields

The electric and magnetic fields produced by an eigenmode $\mathbf{J}_n$ on the surface of the tag are called characteristic fields, and are referred to as ($\mathbf{E}_n$, $\mathbf{H}_n$). One important property of the eigenfields is their orthogonality in the far zone. Based on Poynting's Theorem, the mutual power coupling between the current modes $\mathbf{J}_n$ and $\mathbf{J}_m$ distributed on the surface of the tag can be written as follows

$$\begin{aligned} P_{mn} &= \langle \mathbf{J}_m{}^*, \mathcal{Z}(\mathbf{J}_n) \rangle \\ &= \langle \mathbf{J}_m{}^*, \mathcal{R}(\mathbf{J}_n) \rangle + j\langle \mathbf{J}_m{}^*, \mathcal{X}(\mathbf{J}_n) \rangle \\ &= \int_{S'} \mathbf{E}_m \times \mathbf{H}_n{}^* \cdot ds + j\omega \iiint_{v'} (\mu \mathbf{H}_m \cdot \mathbf{H}_n{}^* - \varepsilon \mathbf{H}_m \cdot \mathbf{H}_n{}^*) dv \end{aligned} \tag{2.118}$$

Here, S' is any surface enclosing the tag and v' is the region enclosed by S'. According to the orthogonality relations in (2.117a, 2.117b, and 2.117c), we have

$$P_{mn} = (1 + j\lambda_n)\delta_{mn} \tag{2.119}$$

If S' is chosen to be a sphere at infinity, then the Eigenfields in the far zone can be expressed by

$$\begin{aligned} \mathbf{E}_n &= -\eta \hat{r} \times \mathbf{H}_n \\ &= \frac{-j\omega\mu}{4\pi r} e^{-jkr} \mathbf{F}_n(\theta, \phi) \end{aligned} \tag{2.120}$$

where $\eta = \sqrt{\mu/\varepsilon}$ is the intrinsic impedance of space, $\hat{r}$ is the unit radial vector perpendicular to S', (θ, φ) are the angular coordinates of the position on S' and $\mathbf{F}_n$ is the pattern of the field. Inserting the far-zone fields in (2.118), the real and imaginary parts of the radiated power can easily be separated as

$$\int_{S'} \mathbf{E}_{\mathrm{m}} \times \mathbf{H}_{\mathrm{n}}^{*} \cdot d\mathbf{s} = \delta_{\mathrm{mn}} \tag{2.121}$$

$$\omega \iiint_{v'} (\mu \mathbf{H}_{\mathrm{m}} \cdot \mathbf{H}_{\mathrm{n}}^{*} - \varepsilon \mathbf{E}_{\mathrm{m}} \cdot \mathbf{E}_{\mathrm{n}}^{*}) dv = \lambda_{\mathrm{n}} \delta_{\mathrm{mn}} \tag{2.122}$$

Relation (2.121) expresses the orthogonality of the eigenfields in the far-zone region. For a single characteristic mode, (2.122) is written by

$$\omega \iiint_{v'} (\mu \mathbf{H}_{\mathrm{n}} \cdot \mathbf{H}_{\mathrm{n}}^{*} - \varepsilon \mathbf{E}_{\mathrm{n}} \cdot \mathbf{E}_{\mathrm{n}}^{*}) dv = \lambda_{\mathrm{n}} \tag{2.123}$$

From (2.123), it is seen that at resonant frequencies where the electric and magnetic energies are equal, the corresponding eigenvalues are zero. At frequencies where $\lambda_{\mathrm{n}} > 0$, the fields are inductive and for $\lambda_{\mathrm{n}} < 0$, the fields are capacitive. According to (2.116), at resonant frequencies, we have

$$\mathrm{X}(\mathbf{J}_{\mathrm{n}}) = 0 \tag{2.124}$$

By applying MoM and converting equation (2.124) into matrix equation, the determinant of the reactance matrix should be zero at the resonant frequencies of the structure in order to have nontrivial solutions.

2.4.3 Modal Solution

The current distribution on the tag can be expanded in terms of the characteristic current modes as

$$\mathbf{J} = \sum_{n=1} a_{\mathrm{n}} \mathbf{J}_{\mathrm{n}} \tag{2.125}$$

where $\mathbf{J}_{\mathrm{n}}$ is the nth characteristic mode and a_{n} is the unknown coefficient in the expansion series. Substituting (2.125) in (2.109) and considering the linearity of the operator, we have

$$\sum_{n} a_{\mathrm{n}} \mathcal{Z}(\mathbf{J}_{\mathrm{n}}) = -\mathbf{E}_{\mathrm{t}}^{\mathrm{inc}} \tag{2.126}$$

By taking an inner product of the two sides of (2.126) with $\mathbf{J}_{\mathrm{m}}$ and using the orthogonality relations in (2.117a, 2.117b, and 2.117c), one can write

$$a_{\mathrm{n}} = -\frac{\langle \mathbf{E}_{\mathrm{t}}^{\mathrm{inc}}, \mathbf{J}_{\mathrm{n}} \rangle}{\langle \mathbf{J}_{\mathrm{n}}, \mathcal{Z}(\mathbf{J}_{\mathrm{n}}) \rangle} = -\frac{\langle \mathbf{E}_{\mathrm{t}}^{\mathrm{inc}}, \mathbf{J}_{\mathrm{n}} \rangle}{1 + j\lambda_{\mathrm{n}}} \tag{2.127}$$

It is seen that the unknown coefficients are strongly dependent on the coupling between the characteristic modes and the incident electric field. By substituting (2.127) in (2.125), the current distribution on the tag is given by

$$\mathbf{J} = -\sum_{n=1} \frac{\left\langle \mathbf{E}_{\mathrm{t}}^{\mathrm{inc}}, \mathbf{J}_{\mathrm{n}} \right\rangle}{1 + j\lambda_n} \mathbf{J}_{\mathrm{n}} = -\sum_{n=1} \frac{V_{\mathrm{n}}}{1 + j\lambda_{\mathrm{n}}} \mathbf{J}_{\mathrm{n}} \tag{2.128}$$

where

$$V_{\mathrm{n}} = \left\langle \mathbf{E}_{\mathrm{t}}^{\mathrm{inc}}, \mathbf{J}_{\mathrm{n}} \right\rangle \tag{2.129}$$

is the coupling coefficient between nth characteristic mode and incident electric field. The electric and magnetic fields scattered from the tag can be written by

$$\mathbf{E} = -\sum_{n=1} \frac{V_{\mathrm{n}}}{1 + j\lambda_{\mathrm{n}}} \mathbf{E}_{\mathrm{n}} \tag{2.130}$$

$$\mathbf{H} = -\sum_{n=1} \frac{V_{\mathrm{n}}}{1 + j\lambda_{\mathrm{n}}} \mathbf{H}_{\mathrm{n}} \tag{2.131}$$

2.4.4 Modal Significance and Characteristic Angle

The variation of eigenvalues, current distribution, and corresponding eigenfields versus frequency provides some useful information about the scattering properties of the tag structure. The modal expansion of the current in (2.128) is inversely dependent on the eigenvalues as

$$MS_{\mathrm{n}} = \left| \frac{1}{1 + j\lambda_{\mathrm{n}}} \right| \tag{2.132}$$

This parameter is called the modal significance. This parameter depends only upon the geometry and dimensions of the tag, and does not vary with the incident excitation. Another parameter, which is very useful in calculating the quality factor of the scatterer at resonant frequencies, is the characteristic angle defined as

$$\alpha_{\mathrm{n}} = 180^\circ - \tan^{-1}(\lambda_{\mathrm{n}}) \tag{2.133}$$

This parameter models the phase angle between a characteristic current, $\mathbf{J}_{\mathrm{n}}$, and the associated characteristic field, $\mathbf{E}_{\mathrm{n}}$. It is clear from (2.132) and (2.133) that at the resonant frequencies of the tag, the characteristic angle is equal to zero and the modal significance has a maximum value of one. These parameters are very useful in calculating the quality factor of the tag at resonant frequencies. As an example, Fig. 2.22 shows the variation of the first two eigenvalues and modal significances

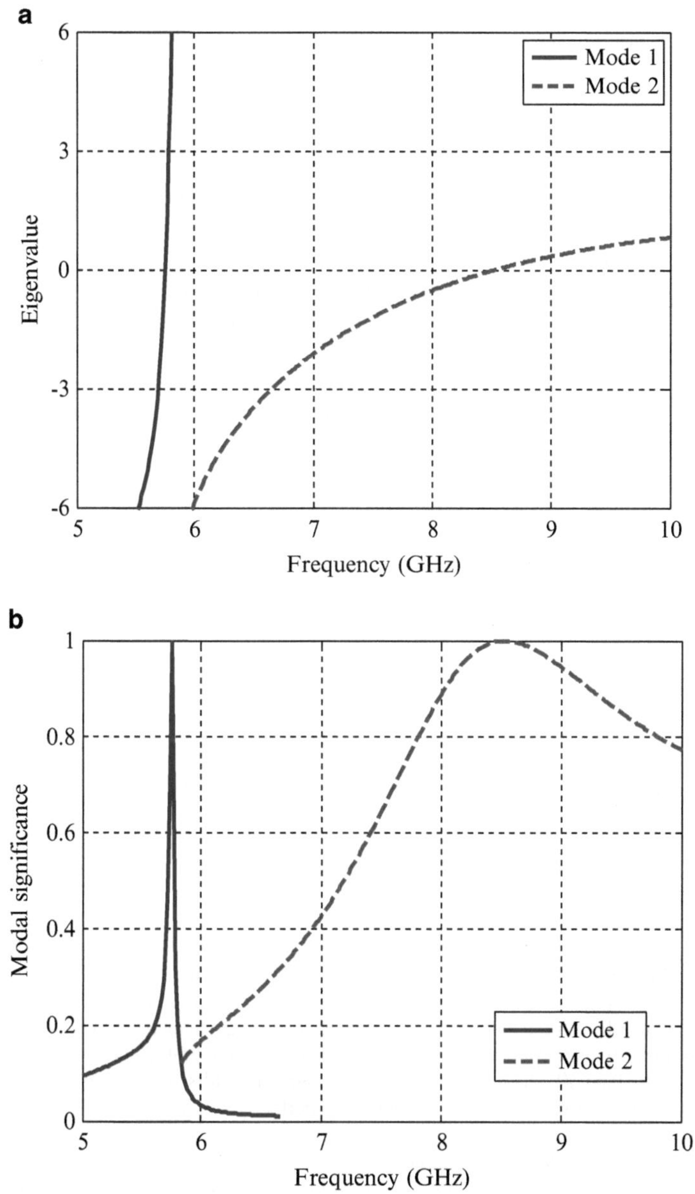

Fig. 2.22 (**a**) Eigenvalues and (**b**) modal significance of the first two characteristic modes of a chipless RFID tag versus frequency

versus frequency for a typical scatterer. In computing the radiating bandwidth of the modes, we need to know the frequencies at which the radiated power is half of that at resonant frequencies. From (2.132), at the frequencies where $\lambda_n = 1$ or $\lambda_n = -1$, the corresponding modal significance is 0.707, and the corresponding characteristic angles are 135° and 225°. Labelling these frequencies f_L and f_H, the quality factor of the characteristic mode at resonant frequency can be calculated from the expression

$$Q \approx \frac{f_0}{f_H - f_L} \quad (2.134)$$

This approximation is only valid for high-Q resonators. There have been numerous formulae for calculating the quality factor of a scatterer. In our application where we try to implement resonators with high quality factor, all the proposed formulae give approximately the same results with slight variations. Another simple formula useful in calculating the quality factor of the resonators embedded on chipless RFID tags is given by the expression

$$Q_n = \frac{\omega}{2} \frac{d\lambda_n}{d\omega} \quad (2.135)$$

which was proposed by Harrington [53].

References

1. Baum CE (2012) The singularity expansion method in electromagnetics: a summary survey and open questions. SUMMA Foundation, Akron, OH
2. Baum CE (1971) On the singularity expansion method for the solution of electromagnetic interaction problems. Interaction Note 88, Air Force Weapons Lab., pp 1–111
3. Tesche F (1973) The far-field response of a step-excited linear antenna using SEM. IEEE Trans Antenna Propag 23(6):834–838
4. Baum CE (1986) The singularity expansion method: background and developments. IEEE Antennas Propag Soc Newslett 28(4):14–22
5. Barnes PR (1972) On the singularity expansion method as applied to the EMP analysis and simulation of the cylindrical dipole antenna. AFWL Interaction Note 146, Apr 1972
6. Umashankar KR, Wilton DR (1972) Transient scattering by a thin wire in free space and above ground plane using the singularity expansion method. AFWL Interaction Note 236, Apr 1972
7. Umashankar KR, Wilton DR (1974) Transient scattering characterization of circular loop using singularity expansion method. AFWL Interaction Note 259, 1974
8. Martinez JP, Pine ZL, Tesche FM (1972) Numerical results of the singularity expansion method as applied to a plane wave incident on a perfectly conducting sphere. AFWL Interaction Note 112, 1972
9. Dolph CL, Cho SK (1980) On the relationship between the singularity expansion method and the mathematical theory of scattering. IEEE Trans Antennas Propag 28(6):888–897
10. Baum CB (1976) Emerging technology for transient and broad-band analysis and synthesis of antennas and scatterers. Proc IEEE 64(11):1598–1616
11. Licul S, Davis WD (2005) Unified frequency and time-domain antenna modeling and characterization. IEEE Trans Antennas Propag 53(9):2882–2888
12. Licul S, Davis WD (2004) Pole/residue modeling of UWB antenna systems. IEEE antennas and propagation society international symposium, 2004, pp 1748–1751
13. Davis WA, Licul S (2004) Ultra-wideband antennas, in Introduction to Ultra-Wideband Communications. Reed JH (eds) Englewood Cliffs, NJ: Prentice-Hall

14. Caratelli D, Yarokoy A (2010) Unified time- and frequency-domain approach for accurate modeling of electromagnetic radiation processes in ultrawideband antennas. IEEE Trans Antennas Propag 58(10):3239–3255
15. Bedrosian G (1977) Stick-model characterization of natural frequencies and natural modes of the aircraft. AFWL Interaction Note 326, 1977
16. Rothwell EJ, Nyquist DP, Chen YF, Drachman B (1985) Radar target discrimination using the excitation-pulse technique. IEEE Trans Antennas Propag 33(9):929–937
17. Li Q, Ilavarasan P, Ross JE, Rothwell EJ (1998) Radar target identification using a combined early-time/late-time E-pulse technique. IEEE Trans Antennas Propag 46(9):1272–1278
18. Chen YF, Nyquist DP, Rothwell EJ, Webb L, Drachman B (1986) Radar target discrimination by convolution of radar return with extinction-pulses and single-mode extraction signals. IEEE Trans Antennas Propag 34(7):896–904
19. Baum CE (1996) Discrimination of buried targets via the singularity expansion. AFWL Interaction Note 521, 1996
20. Chen CC, Peters L Jr, Burnside WD (1995) Ground penetration radar target classification via complex natural resonances. Antennas and Propagation Society International Symposium, pp 1586–1589
21. Chen CC, Higgins MB, O'Neill K, Detsch R (1997) Buried unexploded ordnance identification via complex natural resonances. IEEE Trans Antennas Propag 45(11):1645–1654
22. Blischak A, Manteghi M (2009) Pole residue techniques for chipless RFID detection. In Antennas and Propagation Society International Symposium, 2009. APSURSI '09. IEEE, pp 1–4.
23. Blischak AT, Manteghi M (2011) Embedded singularity chipless RFID tags. IEEE Trans Antennas Propag 59:3961–3968
24. Rezaiesarlak R, Manteghi M (2014) Complex-natural-resonance-based design of chipless RFID tag for high density data. IEEE Trans Antennas Propag 52:3109–3121
25. Hong SK, Davis WA (2013) Use of tumor-specific resonance for more efficient microwave hyperthermia of breast cancer. Microw Opt Technol Lett 55(11):2659–2665
26. Hong SK, Davis WA (2012) Use of tumor-specific resonances in microwave hyperthermia of breast. In Antennas and Propagation Society International Symposium (APSURSI), 2012 IEEE
27. Manteghi M, Cooper DB, Vlachos PP (2012) Application of singularity expansion method for monitoring the deployment of arterial stents. Microw Opt Technol Lett 54(10):2241–2246
28. Van Bladel J (2007) Electromagnetic fields, 2nd edn. IEEE Press, New York
29. Salehi M, Manteghi M (2014) Transient characteristics of small antennas. IEEE Trans Antennas Propag 65:2418–2429
30. Salehi M, Manteghi M (2013) A wideband frequency-shift keying modulation technique using transient state of a small antenna. Prog Electromagnet Res 143:421–445
31. Salehi M, Manteghi M (2014) Self-contained compact transmitter for high-rate transmission. Electron Lett 50(4):313–316
32. Schelkunoff SA (1944) Representation of impedance functions in terms of resonant frequencies. Proc IRE 32(2):83–90
33. Baum CE (1976) The singularity expansion method. In: Transient electromagnetic fields. Springer, New York
34. Jin J-M (2010) Theory and computation of electromagnetic fields. Wiley, Hoboken, NJ
35. Carrier G, Krook M, Pearson C (1966) Functions of a complex variable. McGraw Hill, New York
36. Baum CE, Pearson LW (1981) On the convergence and numerical sensitivity of the SEM pole-series in early-time scattering response. Electromagnetics 1(2):209–228
37. Heyman E, Felsen LB (1985) A wavefront interpretation of the singularity expansion method. IEEE Trans Antennas Propag 37(7):706–718
38. Felsen LB (1984) Progressive and oscillatory waves for hybrid synthesis of source excited propagation and diffraction. IEEE Trans Antennas Propag 32:775–796

39. Heyman E, Felsen LB (1982) Creeping waves and resonances in transient scattering by smooth convex objects. IEEE Trans Antennas Propag 31:426–437
40. Altes RA (1976) Sonar for generalized target description and its similarity to animal echolocation systems. J Acoust Soc Am 59:97–106
41. Li L, Tan AE, Jhamb K, Rambabu K (2013) Characteristics of ultra-wideband pulse scattered from metal planar objects. IEEE Trans Antennas Propag 61(6):3197–3206
42. Tesche FM (1973) On the analysis of scattering and antenna problems using the singularity expansion technique. IEEE Trans Antennas Propag 21(1):53–62
43. Chen-To Chai (1994) Dyadic green function in electromagnetic theory. IEEE press series on electromagnetic waves. IEEE Press, New York
44. Rahmat-samii Y (1975) On the question of computation of the Dyadic Green's function at the source region in waveguides and cavities. Trans Microw Theory Tech 23(9):762–765
45. Baum CE (1975) On the Eigenmode expansion method for electromagnetic scattering and antenna problems, part 1: some basic relations for Eigenmode expansion, and their relation to the singularity expansion. Interaction Note 229, Air Force Weapons Lab., pp 1–94
46. Harrington RF (2001) Time-harmonic electromagnetic fields. Wiley-IEEE press, New York
47. Garbacz RJ, Turpin RH (1971) A generalized expansion for radiation and scattered fields. IEEE Trans Antennas Propag 19(1):348–358
48. Harrington RF, Mautz JR (1971) Theory of characteristic modes for conducting bodies. IEEE Trans Antennas Propag 19(5):622–628
49. Cabedo-Fabres M, Antonino-Daviu E, Valero-Nogueira A, Bataller MF (2007) The theory of characteristic modes revisited: a contribution to the design of antenna for modern applications. IEEE Antennas Propag Mag 49(5):52–68
50. Liu D, Garbacz PJ, Pozar DM (1990) Antenna synthesis and optimization using generalized characteristic modes. IEEE Trans Antennas Propag 38(6):862–868
51. Capek M, Hazdra P, Eicher J (2012) A method for the evaluation of radiation Q based on modal approach. IEEE Trans Antennas Propag 60(10):4556–4567
52. Rezaiesarlak R, Manteghi M (2014) On the application of characteristic modes for the design of chipless RFID tags. APS/URSI 2014, Memphis, TN
53. Harrington RF, Mautz JR (1972) Control of radar scattering by reactive loading. IEEE Trans Antennas Propag 20(4):446–454

Chapter 3
Chipless RFID Tag

3.1 Introduction

One key element of a chipless RFID system is the tag. Since it is chipless, it acts both as the scatterer and encoder. As the scatterer, it needs to reradiate the incident field as much as possible in order to maximize signal-to-noise ratio (SNR) in the reader. As the encoder, it needs to encode a high density of data on the backscattered signal. There are some challenges in attaining all desired characteristics of the tag as the scatterer and encoder altogether. Although many designs have been proposed as chipless RFID tags, they can be categorized into two general groups [1].

In the first group, called time-domain reflectometry-based (TDR) design, the tag includes some discontinuities along a long transmission line. The positions of the discontinuities encode the data by a train of pulses shifted corresponding to the positions of the discontinuities. Surface acoustic wave (SAW) tags are an example of this category. The schematic view of a SAW RFID tag is shown in Fig. 3.1. It includes an antenna, piezoelectric surface, and multiple reflectors which encode the data on the signal [2]. The incident electromagnetic pulse received by the antenna is converted to the acoustic wave through the piezoelectric substrate. The SAW is affected by a number of reflectors, which create a number of shifted pulses corresponding to the positions of the reflectors. Although SAW tags are nowadays used in some commercial applications, there are still some issues to be addressed to make them compatible for RFID applications. Reduction in size and loss, and increase in data capacity and reading range are some of these issues. Additionally, due to the costly process of making the SAW and attaching it to the antenna, this type of RFID tag is more expensive than the silicon-based tags [3]. The same idea was employed by utilizing delay line instead of piezoelectric substrate, incorporated by some discontinuities instead of reflectors [4, 5]. In [6], a chipless RFID tag based on group delay engineered dispersive delay structures is proposed. The tag employs transmission-type all-pass dispersive delay structures (DDSs, shown in

R. Rezaiesarlak, M. Manteghi, *Chipless RFID*, DOI 10.1007/978-3-319-10169-9_3

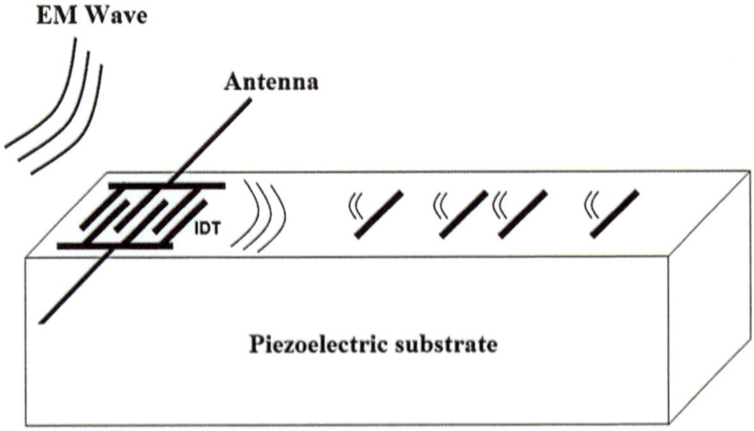

Fig. 3.1 Schematic view of a surface acoustic wave tag

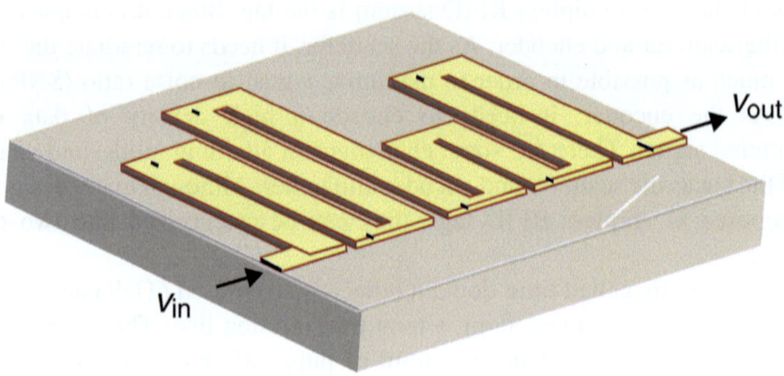

Fig. 3.2 Dispersive delay structures [6] with permission from IEEE

Fig. 3.2) to assign the pulse position modulation (PPM) code onto the interrogating signal. The proposed chipless RFID system based on DDSs is shown in Fig. 3.3. The interrogating signal includes three pulses modulated with three different frequencies. Depending on the frequency of the incident pulse, the structure introduces the required phase shift corresponding to bits 0 or 1. As an example, the pulses can be positioned in the first half or the second half of a bit interval to encode a bit 0 or a bit 1, respectively. In practical applications, TDR-based tags are loaded with an antenna, which increases the size. In addition, the long transmission line included in these structure introduces some loss in the transmission path of the signal.

In the second group of chipless RFID tags, which is called spectral-based tags, the ID of the tag is incorporated into the spectral response of the scattered signal. In these designs, the frequency band of operation is divided into N sections, corresponding to N bits. According to Fig. 3.4, the presence and absence of

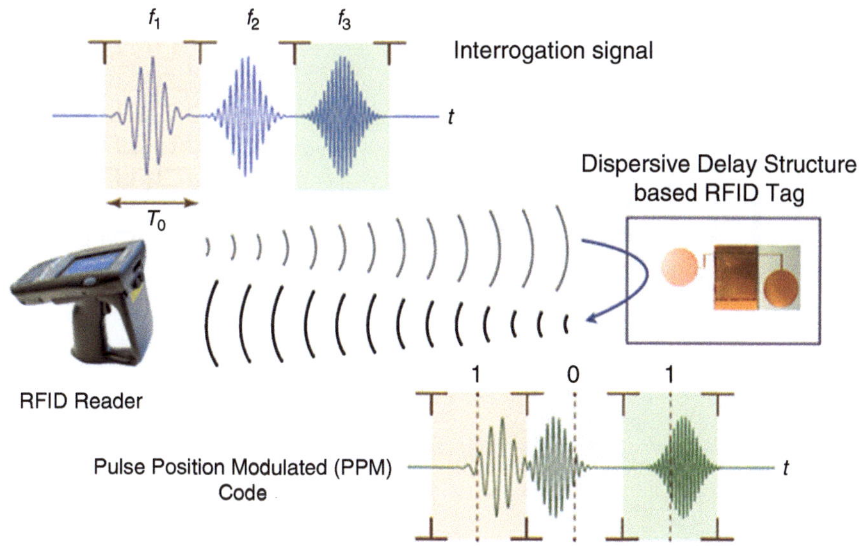

Fig. 3.3 Chipless RFID system based on DDSs [6] with permission from IEEE

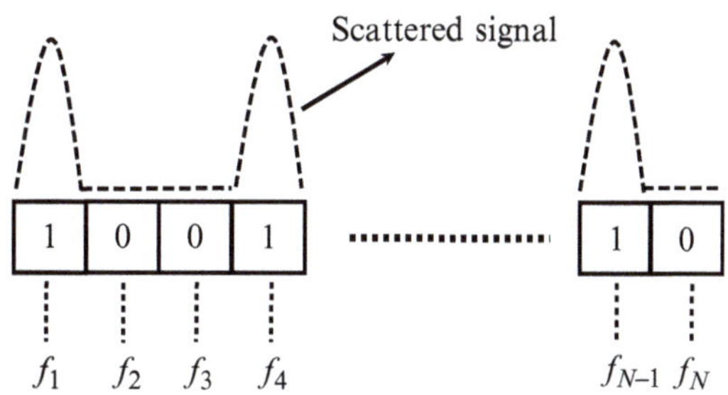

Fig. 3.4 Assigned resonant frequencies for a chipless RFID tag

resonance at each section of the frequency band is associated with bits 1 and 0, respectively. By increasing the number of bits on the tag, the couplings between the resonances are increased. Hence, by removing a specific resonance, the couplings between the other resonators are changed and as a result, the resonant frequencies of the other resonances are altered. Therefore, we need to increase the quality factor of the resonator in order to decrease the coupling between them [7, 8].

The first spectral-based design is shown in Fig. 3.5a. It is an 11-bit tag including 11 dipoles, each corresponding to one bit [9]. In order to decrease the coupling between the dipoles, they are placed far from each other, which increases the size of the tag. Another drawback of the tag is the low quality factor of the dipoles, which is not suitable for high density of data. In 2006, a chipless RFID tag based on the fractal Hilbert curve was proposed [10]. The configuration of the proposed tag is depicted in Fig. 3.5b. From an electromagnetic point of view, such a curve provides

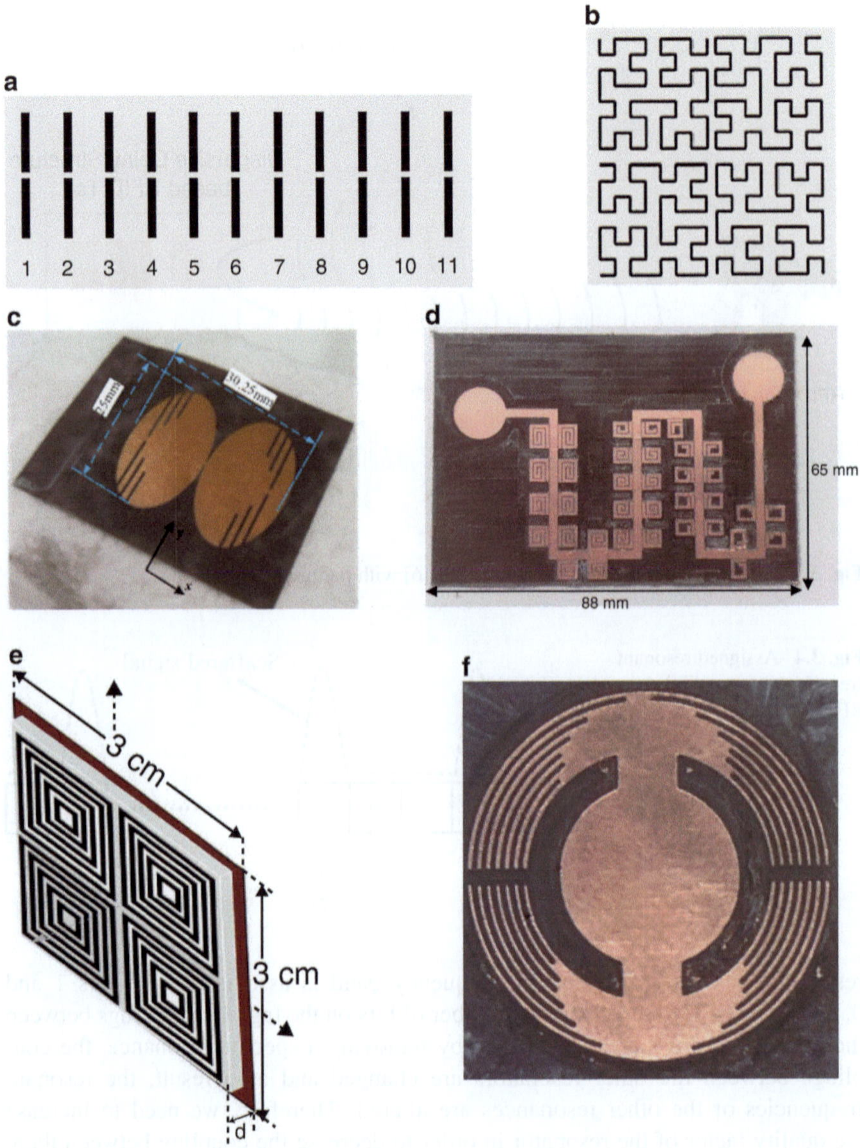

Fig. 3.5 Spectral-based chipless RFID tags using (**a**) dipole resonators [9] with permission from IEEE (**b**) fractal Hilbert curve [10] with permission from IEEE (**c**) slot resonators [3] with permission from IEEE (**d**) Resonant circuitry attached to the transmitting/receiving antennas as a chipless RFID tag [1] with permission from IEEE (**e**) chipless RFID tag based on high-impedance surface [13] with permission from IEEE (**f**) 24-bit tag using quarter-wavelength slot resonators [7] with permission from IEEE

a structure that can resonate at a wavelength much longer than its physical size. For high densities of data where we need to pack the resonances into a limited frequency band, this structure involves some difficulties in the fabrication process. In 2007, a simple structure based on the slot resonator was proposed [11]. By inserting quarter-wavelength slot resonators on a metallic plane as seen in Fig. 3.5c, the ID of the tag is adjusted by the resonant frequencies of the slots. Later on, this structure was used as a basis for designing compact multi-bit chipless RFID tags [7]. Figure 3.5d shows another chipless RFID tag in which a receiving antenna is attached to a resonant circuitry that encodes data on the signal and then transmits the encoded signal through a transmitting antenna [12]. The antennas are placed in different polarizations for transmitting and receiving purposes. Compared to the spectral-based chipless RFID tags in which the structure acts both as the scatterer and encoder, the tag shown in Fig. 3.5d has larger size with higher loss. Hence, this design is not suitable for compact chipless RFID tags. Figure 3.5e shows a chipless RFID tag designed based on high impedance surfaces. In this tag, by employing a multi-resonant HIS unit cell, several bits can be stored in the structure [13]. The states "total reflection" and "total absorption" encode bits 0 and 1, respectively. The ground plane of the microstrip participates in the resonant mechanism of the structure, which makes the tag bulky in some applications. The 24-bit tag represented in Fig. 3.5f contains 24 quarter-wavelength slots, each resonating at the specified frequency [7].

Besides all aforementioned designs proposed as chipless RFID tags, there is a need for a systematic design, which includes all effective structural parameters in the design process. As the encoder, we need to consider the quality factor and resonant frequency tunability of the embedded resonators. As the scatterer, the residue of the poles and radar cross section of the tag and their dependency on polarization and direction should be considered in the design process. In this chapter, two design approaches based on singularity expansion method (SEM) and characteristic mode theory (CMT) are presented. First, complex natural resonance-based design of chipless RFID tags is introduced. By monitoring the effects of structural dimensions on the damping factor and resonant frequency of the resonators, the process of encoding the data onto the tag is presented.

In the following, another design process of the tag based on the theory of characteristic modes is introduced, which provides the resonant frequency, quality factor, and additionally the intensity of the characteristic fields in the far-zone region [14]. Although the design procedure is general and can be used for any arbitrary resonant-based structure, the quarter-wavelength slot resonators is used as the resonant circuitry in the presented design procedure.

This structure exhibits some desired features compared to other proposed resonant structures [7, 15]. Its low profile, ease of fabrication, and lightness are the important features of the proposed tag.

3.2 Complex Natural Resonance-Based Design of Chipless RFID Tags

As mentioned in Chap. 2, the impulse response of the tag can be written as the summation over the CNRs combined with an entire-domain function including the early-time response of the tag as

$$H(s) = \sum_{n=-\infty}^{+\infty} \frac{R_{\mathrm{n}}}{s - s_{\mathrm{n}}} + H_{\mathrm{e}}(s) \tag{3.1}$$

As mentioned before, the residues (R_{n}) and entire-domain function ($H_{\mathrm{e}}(.)$) depend strongly on the polarization and direction of the transmitting and receiving antennas, while the CNRs of the tag (s_{n}) are aspect-independent. Although the series included an infinite number of poles, we consider just N fundamental high-Q CNRs of the tag, which are excited strongly by the incident electric field. A single-bit scheme of the tag is shown in Fig. 3.6 with its structural dimensions.

In order to perceive an intuitive description of the scattering modes, the tag shown in Fig. 3.6 is illuminated by a y-polarized electric field propagating in the x-direction. The current distribution on the tag is illustrated in Fig. 3.7 at different time instances. At $t = 0.02$ ns, the impinging wave hits the leading edge of the tag. The current on the tag at $t = 0.04$ ns is shown in Fig. 3.7b. At this time, the incident wave illuminates part of the tag and as the figure shows the induced current is strongly dependent on the source field. At $t = 0.06$ ns, the incident field crosses through the tag and afterward the current distribution can be written by the summation over the natural modes of the tag. Although each scatterer has infinite

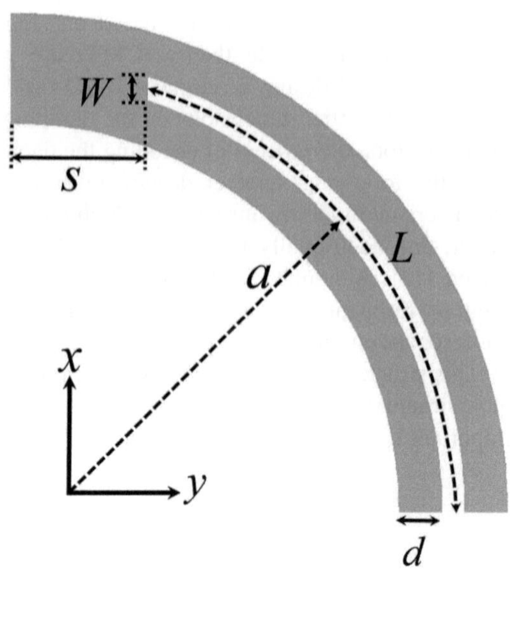

Fig. 3.6 Single-bit tag with structural dimensions [7] with permission from IEEE

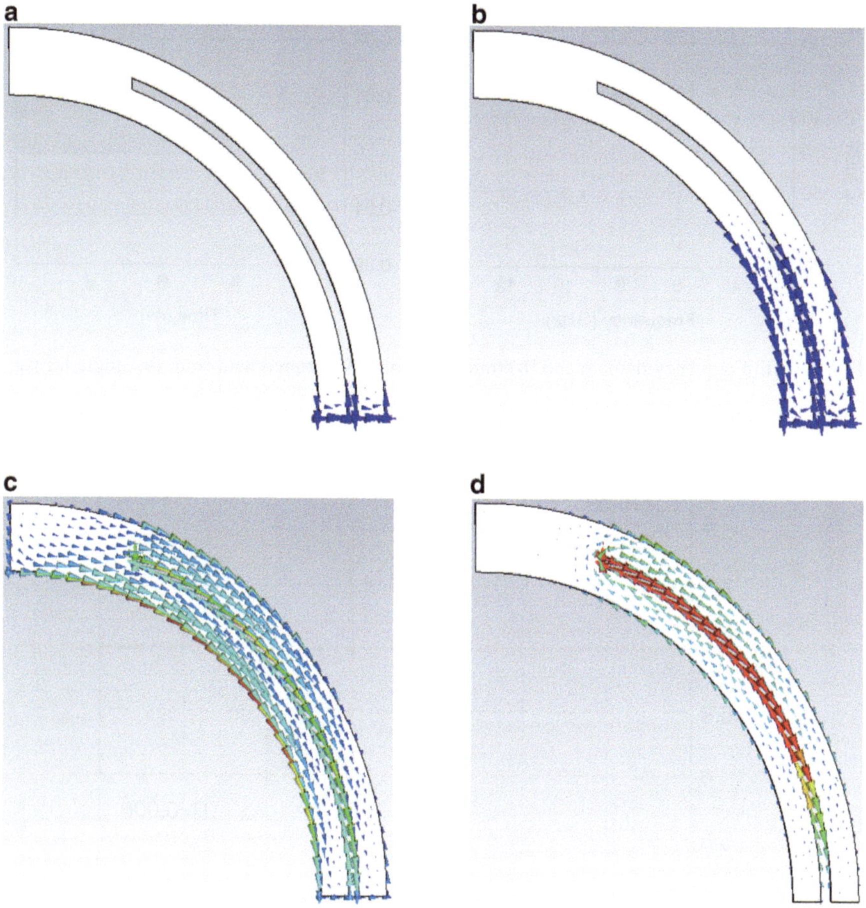

Fig. 3.7 Current distribution on the tag for different time instances. (**a**) t = 0.02 ns, (**b**) t = 0.04 ns, (**c**) t = 0.4 ns, (**d**) t = 0.6ns

CNRs, the dominant resonances excited by the incident field are related to the fundamental resonant frequency of the slot and metal parts. According to the current distribution at $t = 0.4$ ns and $t = 0.6$ ns, it is explicitly seen that the current distribution on the tag is the summation of the natural currents related to the fundamental resonance of the slot and metal. At $t = 0.4$ ns, the current is dominated by the natural modes related to the slot and metal resonant frequencies. At $t = 0.6$ ns and thereafter, the slot's fundamental natural mode is more dominant, because it has a low damping factor compared to other CNRs of the metal.

Assuming $W = 0.3$ mm, $S = 3$ mm, $d = 0.8$ mm, and $L = 12.6$ mm, the time-domain and frequency-domain backscattered field from the tag are depicted in Fig. 3.8. The incident electric field is polarized in the y-direction and propagates

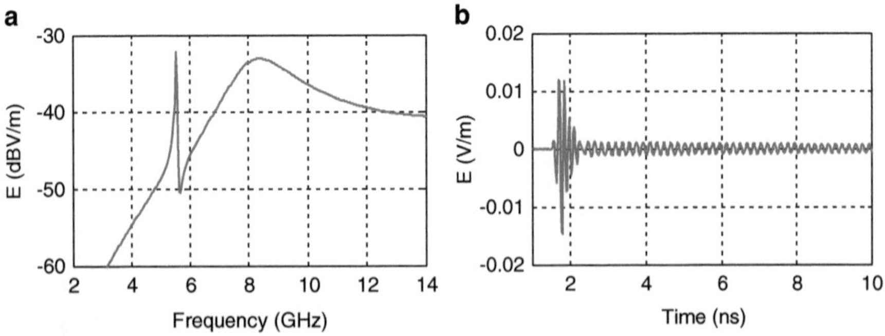

Fig. 3.8 (**a**) Frequency-domain and (**b**) time-domain backscattered field from the single-bit tag

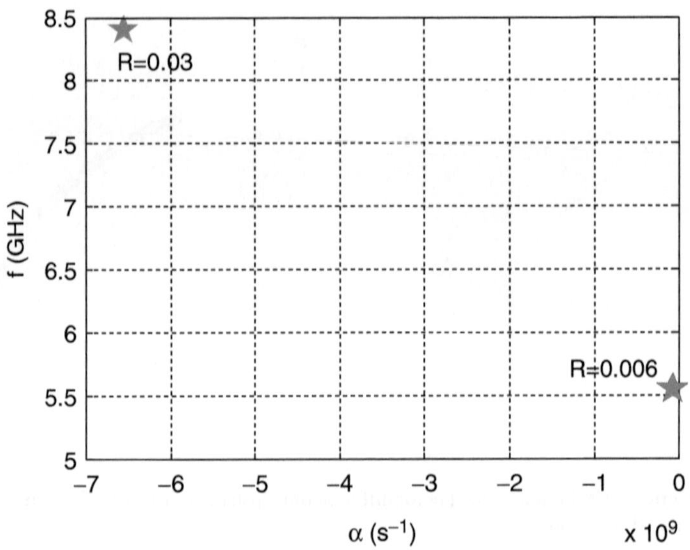

Fig. 3.9 Pole diagram of the single-bit tag

in the z-direction. As the frequency-domain response of the backscattered field shows, two dominantly excited resonant frequencies of the structure are located at $f = 5.54$ GHz and $f = 8.35$ GHz. The pole diagram of the tag is shown in Fig. 3.9. It depicts the resonant frequencies of the tag versus their corresponding damping factors. The residues of the poles are depicted beside them. The high-Q resonance at $f = 5.54$ GHz is related to the fundamental frequency of the slot resonator, and the low-Q resonant frequency at $f = 8.35$ GHz is associated with the metallic part of the tag. Although the CNR of the slot has a lower damping factor than the CNR of the metal, its residue is five times weaker. Since the fundamental resonant frequency of the slot is used in encoding the data onto the tag, the resonant

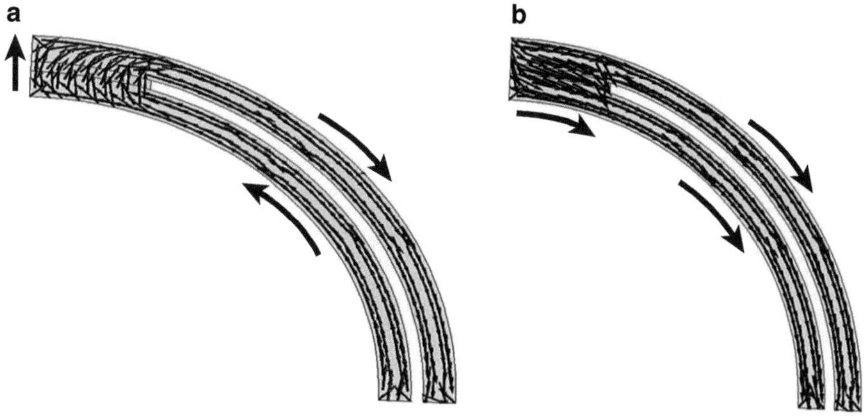

Fig. 3.10 Current distribution on the tag at (**a**) $f = 5.54$ GHz and (**b**) 8.8 GHz [7] with permission from IEEE

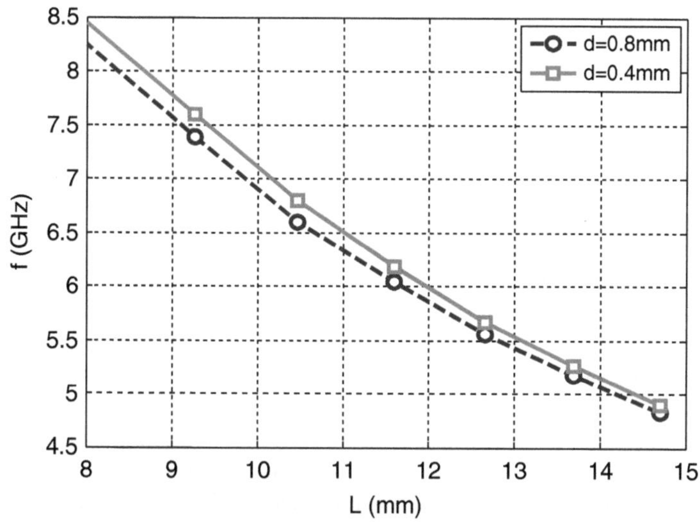

Fig. 3.11 Resonant frequency of the slot versus L. W = 0.3 mm [7] with permission from IEEE

frequency of the metal should be considered in the design procedure and distinguished in the applied detection technique.

In Fig. 3.10, the current distribution on the tag is shown for two resonant frequencies. As it shows, at the resonant frequency of the slot, the current on the arms of the slot are in opposite directions while at $f = 8.35$ GHz, which corresponds to the half-wavelength resonance of the metal, the current on the arms is mostly in the same direction. Based on these current distributions, it can be inferred that the resonant frequency of the slot is mostly sensitive to the slot length L and that the resonant frequency of the metal can be changed by $L+S$. In practical applications, the fundamental CNR of the slot rather than the metal is used for encoding the data onto the tag because of its higher

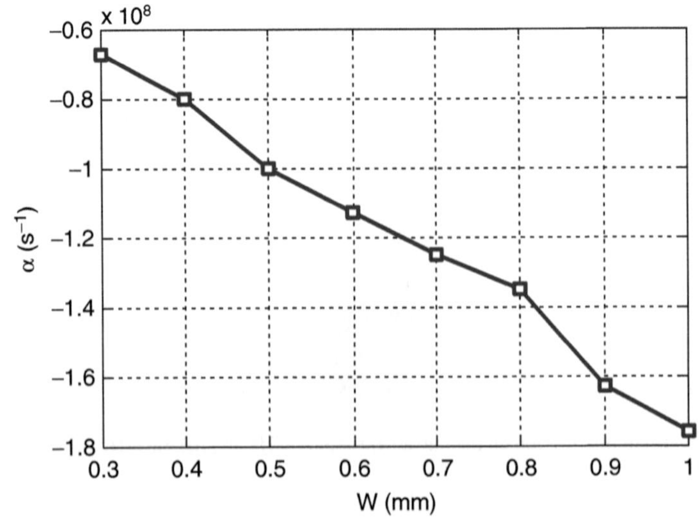

Fig. 3.12 Damping factor of the CNR of the slot versus d for different values of *a*. W = 0.3 mm [7] with permission from IEEE

quality factor. In Fig. 3.11, the resonant frequency of the slot is plotted versus slot length for two different values of *d*. As can be seen, increasing the slot length shifts down its resonant frequency. In addition, for larger values of *d*, as a result of slightly increasing the electrical length of the slot, its resonant frequency decreases.

Therefore, the resonant frequency can be easily adjusted by the slot length (L). Another important parameter to be considered in the design procedure is the damping factor of the resonances. In order to incorporate more resonant frequencies into a relatively narrow frequency band, it is necessary to design the poles with low damping factors or equivalently high quality factors. In Fig. 3.12, the damping factor of the slot's fundamental CNR is plotted versus the width of slot (W). As it shows, by increasing the slot width and consequently increasing the parasitic fields at the open-ended edge of the slot, its damping factor slightly increases. Due to the limitation in prototyping of slots with widths below 0.2 mm, this parameter cannot be used effectively to lower the damping factor of the slot resonator. Assuming slot width of $W = 0.2$ mm, the damping factor of the slot is depicted in Fig. 3.13 in terms of d. As it shows, by decreasing d, the damping factor is significantly decreased. According to the current distribution on the tag as seen in Fig. 3.10a, by decreasing d, the currents on the edges of the resonator are placed closer to each other and as a result, backscattered radiation decreases drastically. By reducing the radiated power, according to the definition of the quality factor

$$Q = \omega \frac{E_{\text{stored}}}{P_{\text{radiation}}} \tag{3.2}$$

the quality factor of the resonator increases which leads to a lower damping factor at that resonant frequency. The quantities ω, E_{stored}, and $P_{\text{radiation}}$ are the radian

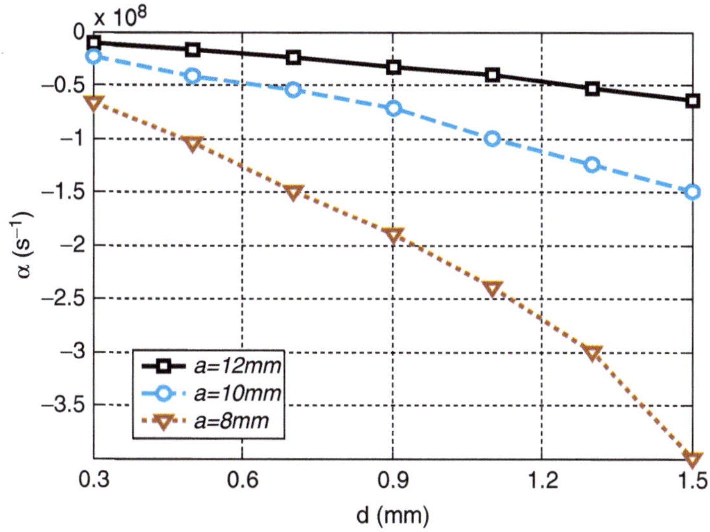

Fig. 3.13 Damping factor of the CNR of the slot versus *d* for different values of *a*. W = 0.3 mm [7] with permission from IEEE

frequency, stored energy around the tag, and radiation power from the tag, respectively. By decreasing the radiation fields the radar cross-section (RCS) of the tag is decreased, which makes the identification process of the tag more challenging especially in the presence of noise and clutter. Assuming a single-bit tag as seen in Fig. 3.6, the late-time response corresponding to fundamental resonance can be written by

$$s(t) = \mathrm{R}e^{-\alpha_0 t}\cos\left(\omega_0 t + \varphi_0\right) \tag{3.3}$$

Where R_0 and φ_0 are the amplitude and phase of the signal resonating at the radian resonant frequency of ω_0 and attenuating with damping factor α_0. The energy of the signal is defined by

$$\begin{aligned} E_s &= \int_{t=0}^{+\infty} s^2(t)dt \\ &= \frac{R^2}{4\alpha_0} + \frac{R^2\alpha_0}{4(\alpha_0{}^2 + \omega_0{}^2)} e^{\frac{2\alpha_0\phi_0}{\omega_0}} \end{aligned} \tag{3.4}$$

For the CNRs with low damping factor as $\omega_0 \gg \alpha$, assuming $\varphi_0 = 0$, the energy of the signal in (3.4) is summarized as

$$E_s = \frac{R^2}{4\alpha_0} \tag{3.5}$$

It is seen that the energy of the single-pole signal is proportional to the square of its amplitude and inverse of its damping factor, when its damping factor is much

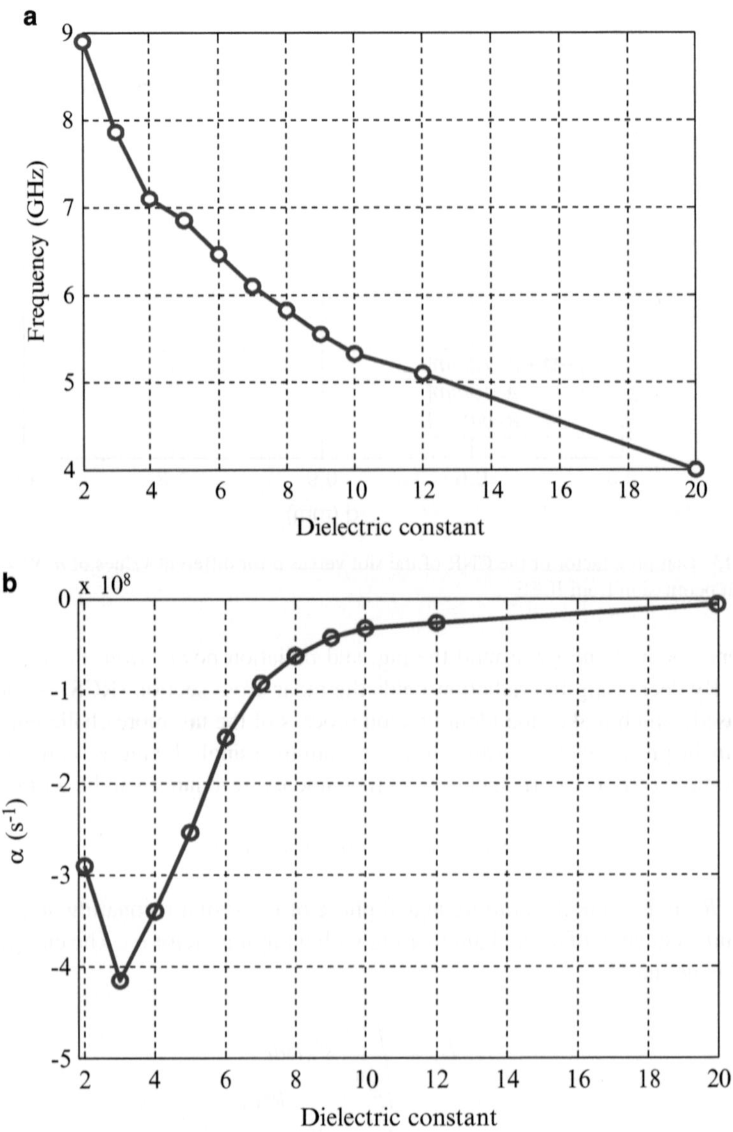

Fig. 3.14 (**a**) Resonant frequency and (**b**) damping factor of the CNR of the tag versus dielectric constant. $a = 10$ mm, $d = 0.8$ mm, $t = 0.3$ mm

smaller than radial frequency. For this reason, it is crucial in the identification process of the scattered signal from chipless RFID tag to use resonances with low damping factors.

In practical application, the tag structure is usually designed on a thin dielectric substrate. Because of the existence of the lossy dielectric, both the resonant frequency and damping factor of the CNRs change. In Fig. 3.14, the variation of

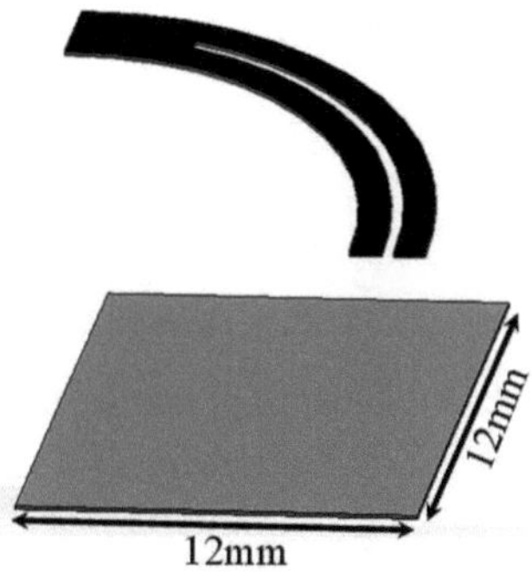

Fig. 3.15 Single-bit tag above a metallic plate

the CNRs versus dielectric constant of the substrate is shown. The thickness of the substrate is assumed 0.7874 mm and the dimensions of the structure are $a = 10$ mm, $d = 0.8$ mm, $W = 0.3$ mm. As it shows by increasing the dielectric constant of the lossless substrate, the resonant frequency and the damping factor of the CNR decrease. Based on the scaling relationship, when a tag is located in a lossy dielectric with dielectric constant ε and conductivity δ [16],

$$s_{\mathrm{n}} = -\frac{\delta}{2\varepsilon} + \sqrt{\left(\frac{\delta}{2\varepsilon}\right)^2 + \left(\frac{\left(s_{\mathrm{n}}^{\mathrm{fs}}\right)^2}{\varepsilon}\right)} \tag{3.6}$$

where $s_{\mathrm{n}}^{\mathrm{fs}}$ and s_{n} are the nth-CNR of the tag in free space and in the lossy dielectric. In the case where the tag is located on the dielectric substrate from one side, neglecting the loss of the dielectric, we have

$$s_{\mathrm{n}} = \frac{s_{\mathrm{n}}^{\mathrm{fs}}}{\sqrt{\varepsilon_{\mathrm{eff}}}} \tag{3.7}$$

According to (3.7), the damping factor and resonant frequency of the tag attached to the dielectric substrate decrease by increasing the dielectric constant of the substrate (according to Figs. 3.13 and 3.14). As the Fig. 3.14 shows, when the dielectric constant of the substrate is between 2 and 3, the damping factor of the CNR increases. In this region, the resonant frequency of the slot is close to the resonant frequency of the metal. Therefore, the coupling effect between these two resonant mechanisms increases the damping factor of the slot's CNR.

Another important parameter in chipless RFID tags is the sensitivity of the CNRs to background objects in the environment. For this purpose, a single-bit chipless RFID tag is assumed above a metallic plate as the Fig. 3.15 shows. The percentage variations of the resonant frequency and damping factor in terms of the distance between tag and plate are illustrated in Fig. 3.16. According to the figures, as a result of coupling, the damping factor of the pole is much more sensitive to the environmental objects than the resonant frequency. This is the reason why the damping factor of the poles is not usually used in the identification process.

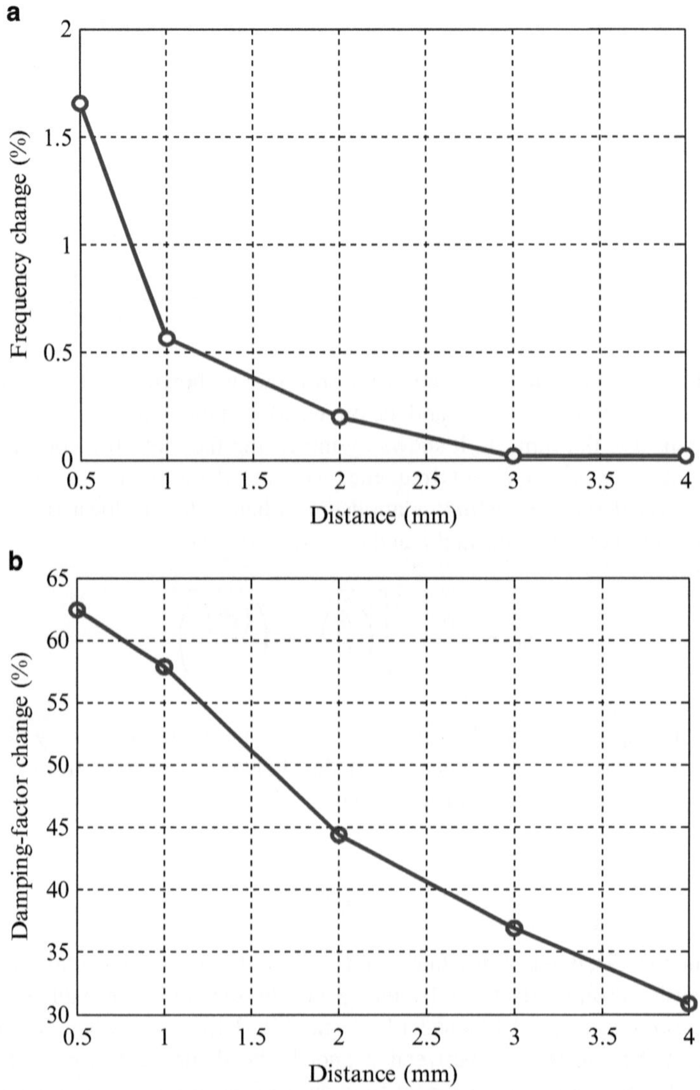

Fig. 3.16 Percentage variations in the (**a**) resonant frequency and (**b**) damping factor of the CNR of the tag versus distant to the metallic plate. $a = 10$ mm, $d = 0.8$ mm, $t = 0.3$ mm

The single-bit tag seen in Fig. 3.6 can be used in the design of multi-bit chipless RFID tags. The coupling between resonances plays a critical role in the design of chipless RFID tags. The presence and absence of a resonance at a specific resonant frequency represents bit 1 or 0, respectively. Therefore, the structure should be designed in such a way that by nulling one resonant frequency, the resonant frequencies of the other resonators do not change.

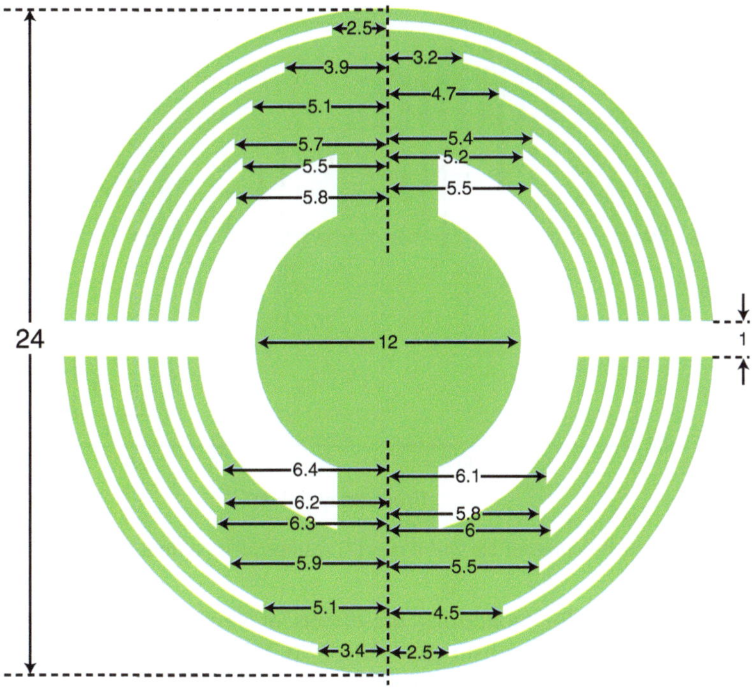

Fig. 3.17 Schematic view of the 24-bit tag [7] with permission from IEEE

A 24-bit tag shown in Fig. 3.17 is designed based on the quarter-wavelength slot resonators on a Rogers RT/Duroid®/5870 ($\varepsilon_r = 2.2$) with a thickness of 0.78 mm. The lengths of the slots are tuned in order to adjust the resonant frequencies of the slots according to the ID of the tag. By filling the slot surface with metal or inserting some stubs along the slot, the corresponding bit is nulled. Therefore, 2^{24} tags with unique IDs can be encoded with 24 slots. As an example, in the second tag, the second and fifth bits of the tag are nulled by inserting some stubs along the slot. The radar cross-sections of the tags are depicted in Fig. 3.18. It is seen that the second and fifth bits of the tag are nulled without any considerable change in the positions of the other resonances.

3.3 Design of Chipless RFID Tag Based on Characteristic Mode Theory

In the previous section, the effects of various structural dimensions of the tag on the CNRs were studied. The scattered field radiated from the natural current mode depends strongly on its distribution on the tag surface. We need to calculate the CNRs and corresponding natural modes on the tag by employing some numerical techniques such as method of moment (MoM).

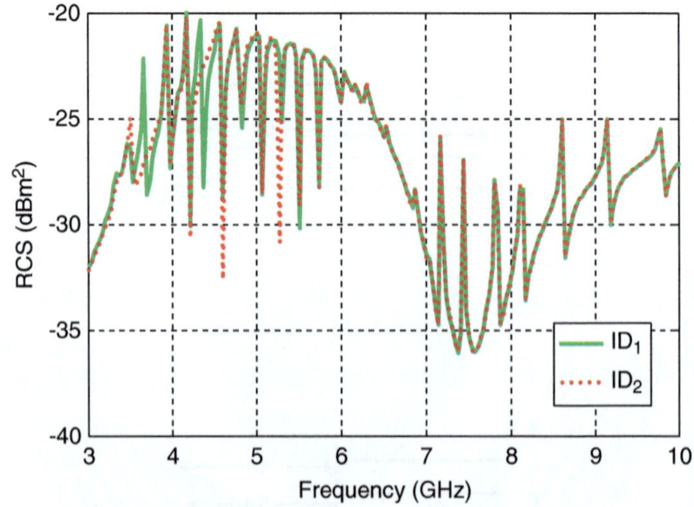

Fig. 3.18 Radar cross-sections of the 24-bit tags [7] with permission from IEEE

Another approach for monitoring the effects of structural parameters on the scattered response is characteristic mode theory (CMT). As mentioned in Chap. 2, by decomposing the current distribution on the tag into its characteristic modes, the resonant and radiation characteristics of the tag can be studied easily at each frequency. In some commercial software such as FEKO, the characteristic modes of the structure and the variations of the eigenvalues, modal significances, and radiated power can be easily monitored versus frequency. This insight is useful in the design of chipless RFID tags.

A single-bit tag with the dimensions shown in Fig. 3.19 is considered. Two parameters d_1 and d_2 are shown in Fig. 3.19, which are initialized at the y-axis in order to study the effects of the metal and slot resonances more accurately. The eigenvalues of the first two characteristic modes of the tag are illustrated in Fig. 3.20 versus frequency for $d = .8$ mm, $W = 0.3$ mm, $d_1 = 3.5$ mm, $d_2 = 0$, $a = 10$ mm, and $L = 12.1$ mm. The resonant frequencies of the tag are the frequencies at which the eigenvalues are equal to zero. In this case, the resonant frequencies of the tag in the 3–10 GHz band are located at $f_1 = 5.7$ GHz and $f_2 = 8.6$ GHz. Figure 3.21 illustrates the modal significances of the characteristic modes, which are calculated from

$$MS_{\mathrm{n}} = \left| \frac{1}{1 + j\lambda_{\mathrm{n}}} \right| \tag{3.8}$$

According to Fig. 3.21, the first resonance of the tag has a much higher quality factor than the second. For high-Q resonances which are usually used in the design of chipless RFID tags, the quality factor of the CNRs can be calculated from the

Fig. 3.19 Single-bit tag illuminated by incident plane wave

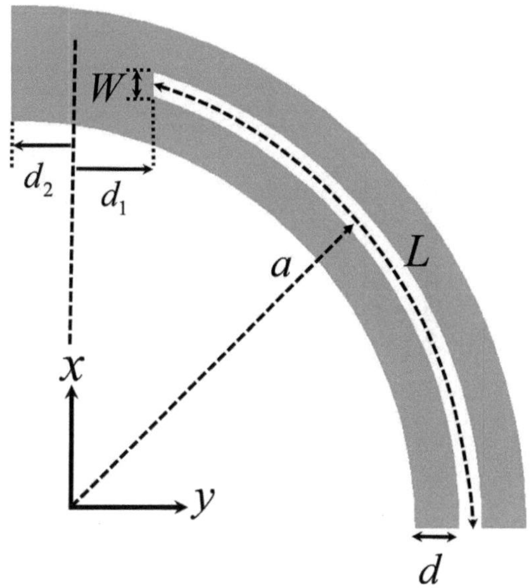

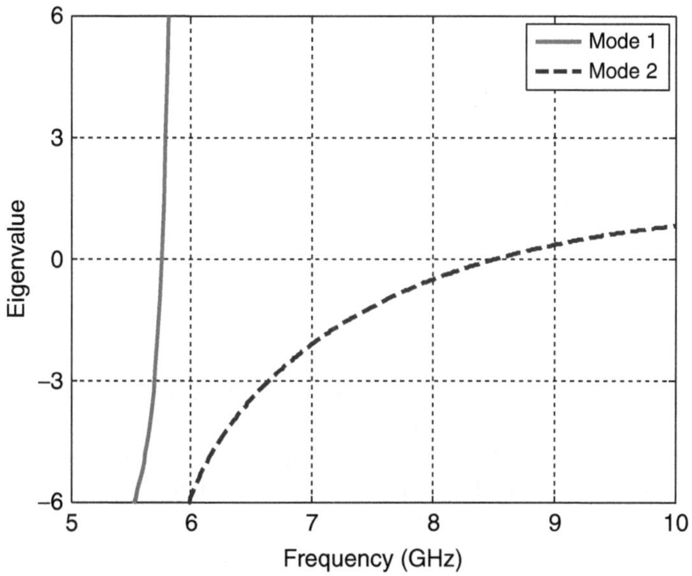

Fig. 3.20 Eigenvalues of the characteristic modes versus frequency

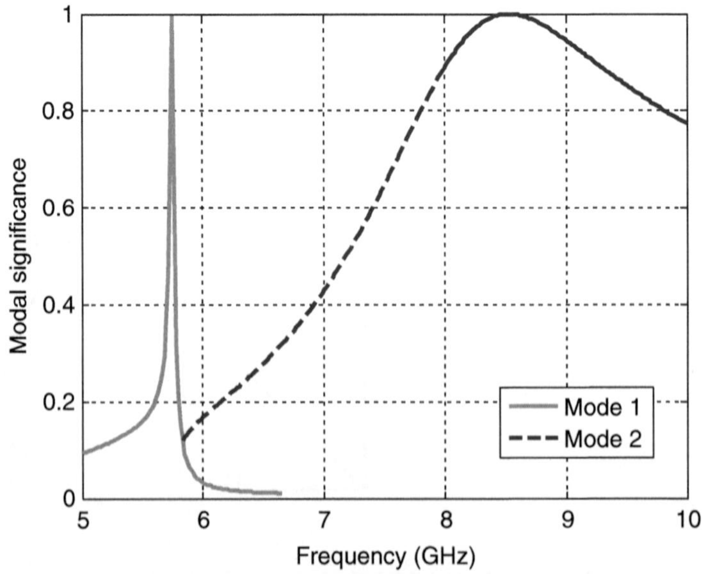

Fig. 3.21 Modal significances of the characteristic modes versus frequency

modal significance (MS) of the modes. Assuming f_L and f_H as the frequencies at which the MS is 0.707, the quality factor can be calculated from

$$Q \approx \frac{f_0}{f_H - f_L} \tag{3.9}$$

where f_0 is the resonant frequency of the tag. The quality factor of the resonances can also be calculated from the characteristic angles of the modes. The characteristic angle of a characteristic mode is obtained from

$$\alpha_n = 180^\circ - \tan^{-1}(\lambda_n) \tag{3.10}$$

As can be seen from (3.10), the characteristic angle is 180° at the resonant frequency and is 135° and 225° at f_L and f_H, respectively. The characteristic angles of the modes are shown in Fig. 3.22 versus frequency. The first resonant frequency at $f_1 = 5.7$ GHz is corresponding to the quarter-wavelength resonant of the slot and the resonant frequency at $f_2 = 8.6$ GHz is related to the half-wavelength resonance of the metal.

The first two characteristic modes of the tag at two resonant frequencies are depicted in Fig. 3.23. As the figures show, the currents on the arms of the slots oppose each other at $f = 5.7$ GHz which is associated to the slot's resonance. The first characteristic mode of the tag at $f = 8.6$ GHz agrees with the current distribution at the half-wavelength resonance of the tag, while it is not the same for the

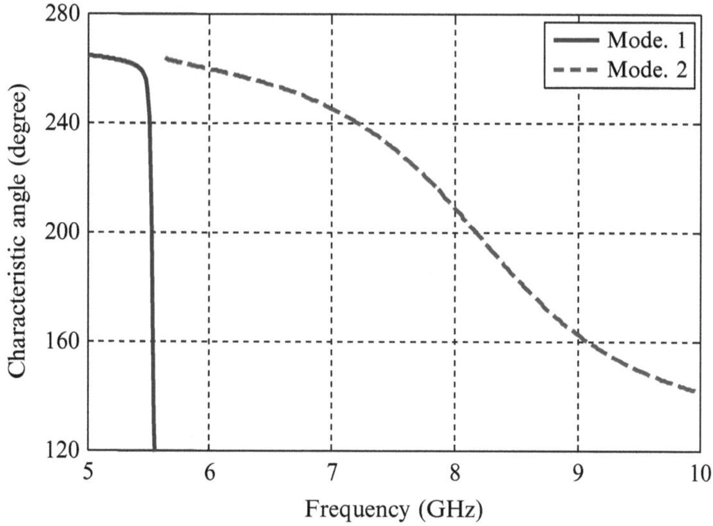

Fig. 3.22 Characteristic angle of the characteristic modes versus frequency

second characteristic mode. The actual current on the tag is the superposition of the characteristic modes weighted with coefficients proportional to the coupling coefficients and modal significances.

Taking advantage of the characteristic mode theory (CMT) in the design procedure, the variations of the resonant frequency and quality factor of the CNRs and field intensity can be monitored easily. According to the current modes, it is seen that the resonant frequency of the slot is dependent strongly upon the slot length and the resonance of the metal is proportional to the metal length. Knowing this, two parameters d_1 and d_2 shown in Fig. 3.19 are used to change the lengths of the slot and metal on the tag structure. In Fig. 3.24, the variations of the first and second resonant frequencies of the tag are shown versus d_1. It is seen that by increasing d_1 and consequently decreasing the slot length, its resonant frequency increases without considerable change in the resonant frequency of the metal. Figure 3.25 shows the variations of the resonant frequencies of the slot and metal versus d_2. By altering d_2, the resonant frequency of the metal changes without considerable variations in the resonant frequency of the slot. It is very useful in the design of the tag to be able to tune the resonant frequencies of the structure separately. Since the resonance of the slot is usually utilized for encoding the ID onto the tag, it is necessary to place the resonance of the metal outside the frequency band of operation or to distinguish it from the slot resonances in the detection process.

When the resonant frequency of the slot is located close to the resonant frequency of the metal, by increasing the coupling between these two resonant frequencies the quality factor of the slot resonator decreases. In Fig. 3.26, the backscattered response from the single-bit tag with dimensions $d_2 = 6$ mm, $a = 12$ mm, $d = 0.8$ mm, $W = 0.3$ mm is shown for three different values of d_1.

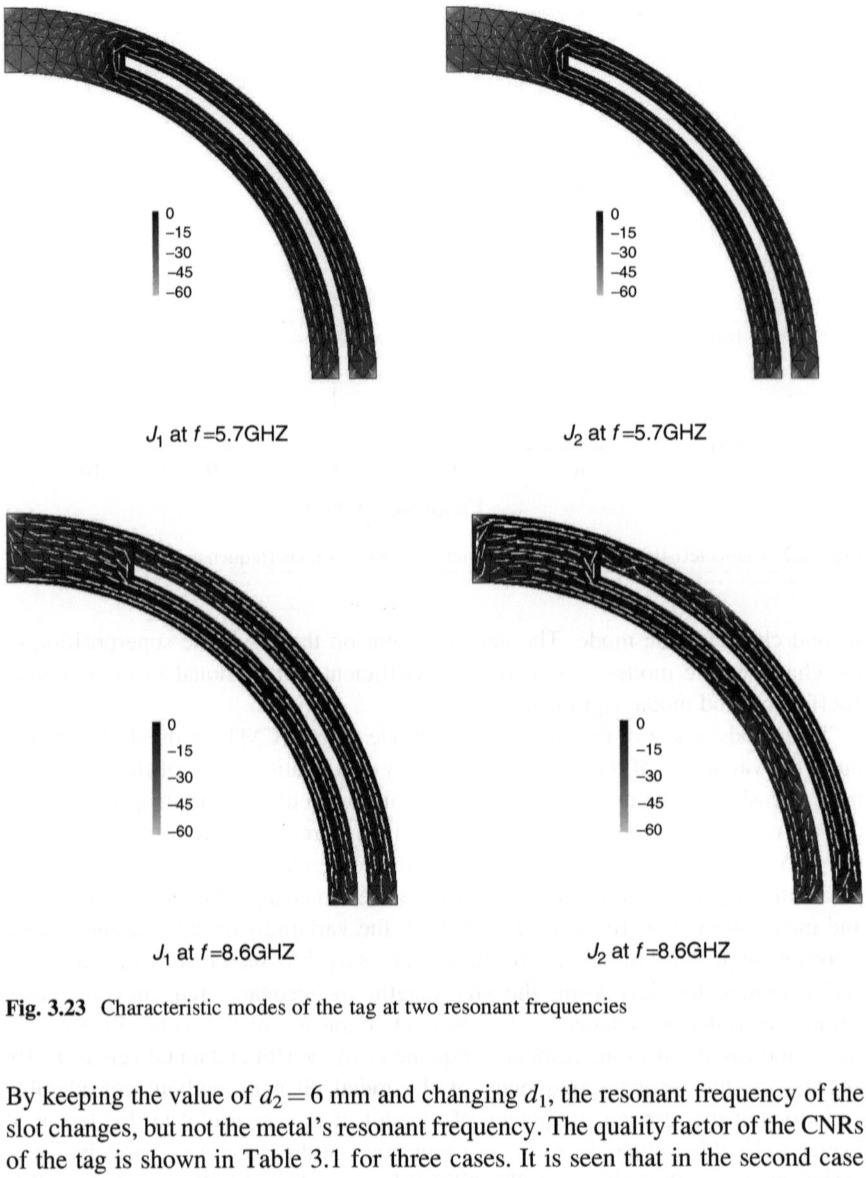

Fig. 3.23 Characteristic modes of the tag at two resonant frequencies

By keeping the value of $d_2 = 6$ mm and changing d_1, the resonant frequency of the slot changes, but not the metal's resonant frequency. The quality factor of the CNRs of the tag is shown in Table 3.1 for three cases. It is seen that in the second case where the resonant frequencies are located in close proximity to each other, the quality factor of the slot's CNR decreases more than ten times. The quantity R in the table is the residue of the corresponding CNR of the tag in the late time. In the second case, the slot's CNR has higher residue, which is very important in the detection process. Based on (3.3), the energy of the CNR is directly proportional to the square of the residue and inversely to the damping factor of the CNR. In Chap. 5, the effects of residue and damping factor on detection of a signal in the presence of noise will be shown. In some applications such as chipless RFID

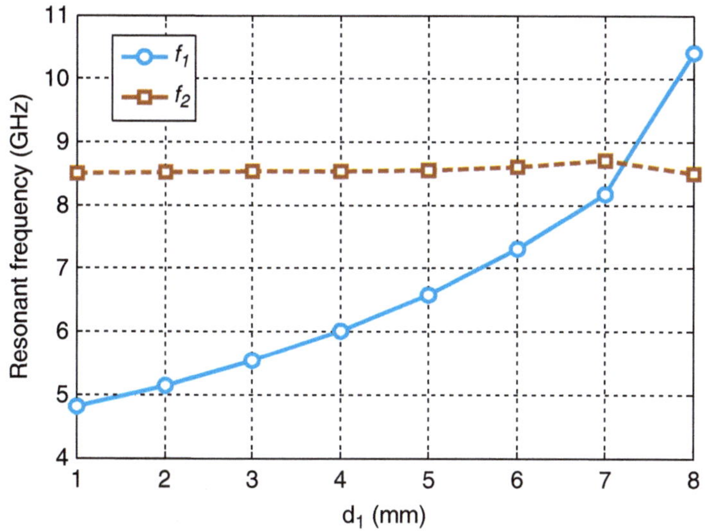

Fig. 3.24 Variation of the resonant frequencies of the tag versus d_1

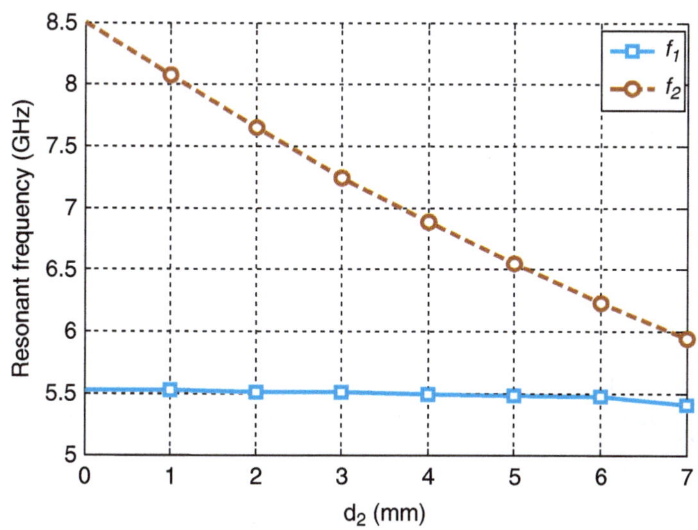

Fig. 3.25 Variation of the resonant frequencies of the tag versus d_2

sensors where few bits are used in the sensing process in the lossy media, the strength of the late-time response is critical in the detection of a signal. In tags with high density of data, a low damping factor is desirable, decreasing the coupling between resonators.

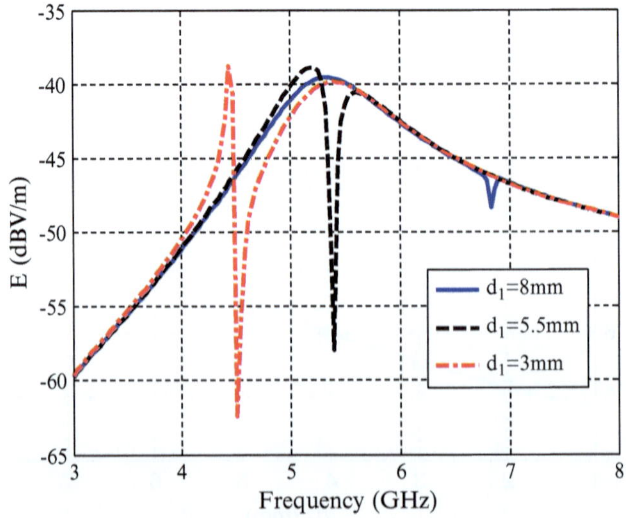

Fig. 3.26 Scattered far-field electric field radiated from the tag. $d_2 = 6$ mm, $a = 12$ mm, $d = 0.8$ mm, $W = 0.3$ mm

Table 3.1 The resonant frequency, quality factor, and residue of the CNR of the slot for different cases

d_1(mm)	Q_{slot}	Q_{metal}	(R)	f_{slot}(GHz)	f_{metal}(GHz)
8	115	4.12	1.2	6.8	5.4
5.5	11.3	6	5	5.2	5.7
3	446	4.12	0.2	4.4	5.4

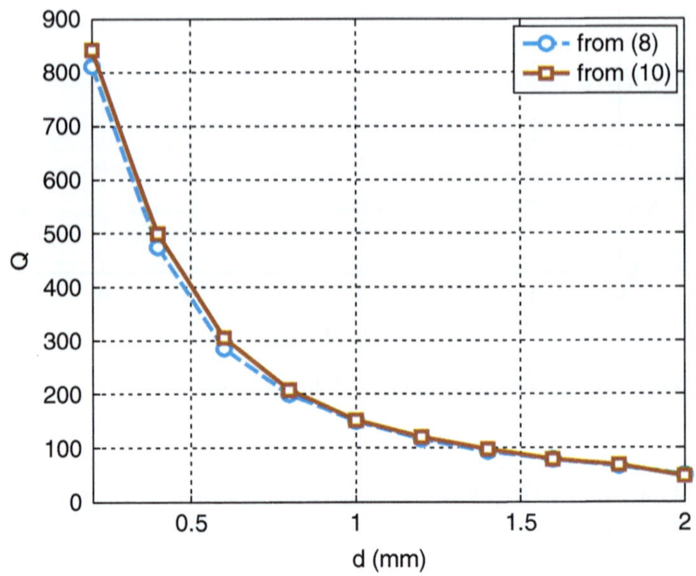

Fig. 3.27 Quality factor of the CNR of the tag versus d

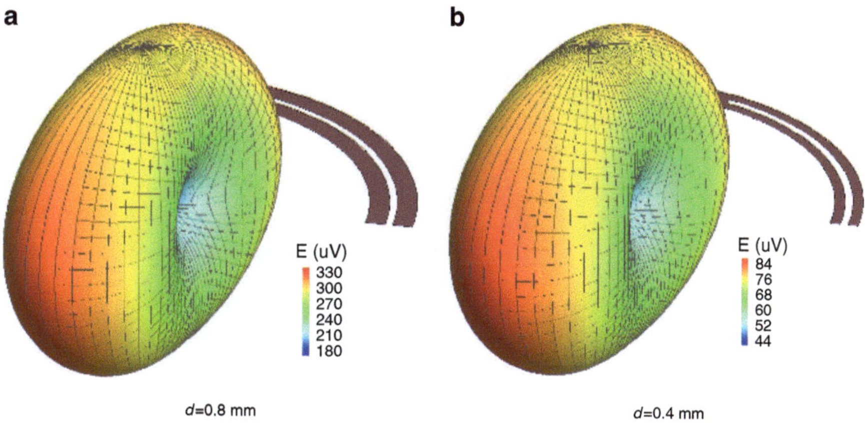

Fig. 3.28 Far-field electric fields radiated from the tag for (**a**) $d = 0.8$ mm and (**b**) $d = 0.4$ mm

As mentioned before, the quality factor of the resonances of the tag can be controlled by d. The variation of the quality factor of the slot is shown in Fig. 3.27 versus d. By increasing the arm width d, the quality factor of the slot's resonance decreases, which agrees with the discussion in Sect. 3.2. By decreasing d and increasing the quality factor of the CNR, more energy is localized in the reactive near-field of the tag, which leads to a decrease in the radiation from the tag. Therefore, the RCS of the tag decreases. As an example, the far-field radiation pattern of the tag is seen in Fig. 3.28 for $d = 0.8$ mm and $d = 0.4$ mm. The maximum radiation intensity of the far-zone electric field is 330 μV/m for $d = 0.8$ mm, compared to 84 μV/m for $d = 0.4$ mm.

By following the aforementioned design procedure and monitoring the effects of structural dimensions on the resonant frequency, quality factor, and intensity of the backscattered field from the tag, two 4-bit tags are designed. The schematic view of the tag is shown in Fig. 3.29. The slot resonators embedded on the metallic plane are used for encoding the data. Two different cases of $d_2 = 3$ mm and $d_2 = 7$ mm are considered. By illuminating the tag with an incident plane wave polarized in the x-direction and propagating in the z-direction, the backscattered response from the tag is shown in Fig. 3.30 for two values of d_2. The pole diagram of the tag is depicted in Fig. 3.31 for two different cases. The resonance of the metal is located out of the frequency band of the slot's resonances for $d_2 = 3$ mm. By increasing d_2 sufficiently, the resonance of the metal can be within the frequency band of the slot's resonances. In this case, the damping factor of the slot's resonances increases as a result of coupling, which leads to a lower quality factor. This must be avoided in the design of the tags with high density of data in order to pack many resonant frequencies in a narrow frequency band.

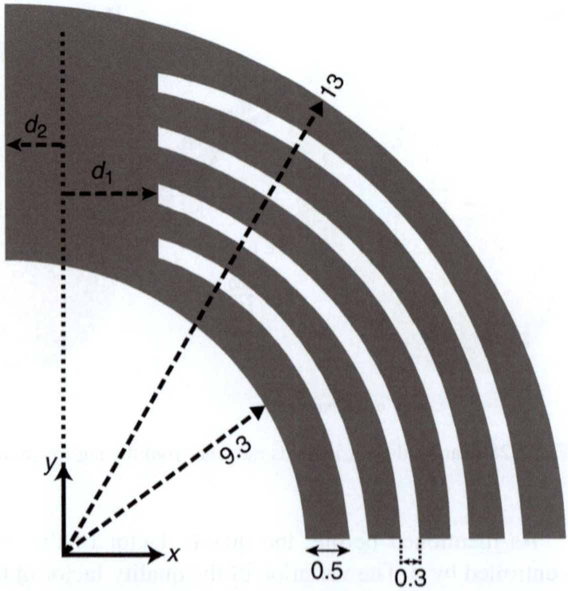

Fig. 3.29 Schematic view of the designed 4-bit tag. Units: mm

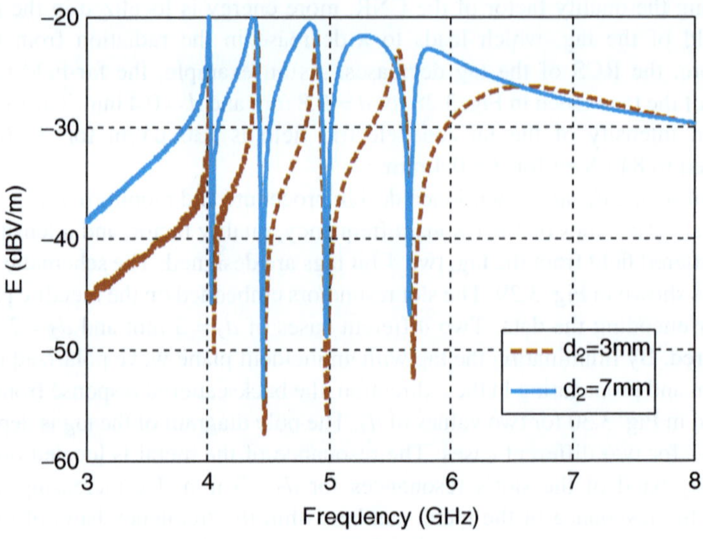

Fig. 3.30 The simulated backscattered electric field from 4-bit tags

Two fabricated tags are shown in Fig. 3.32. The tags were designed on a Rogers RT/Duroid®/3003 ($\varepsilon_r = 3$) with a thickness of 0.7874 mm. Assuming the incident electric field directed in x and propagating in z, the measured backscattered electric field from the tags is depicted in Fig. 3.33. Although the data are incorporated as

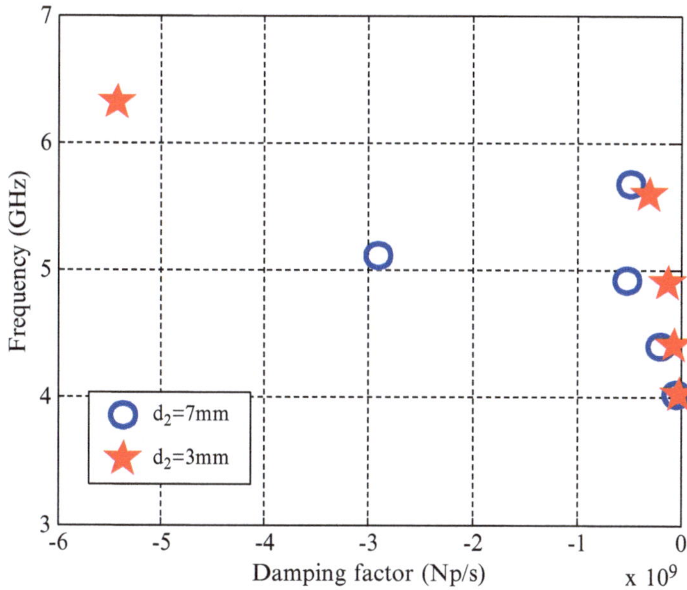

Fig. 3.31 Pole diagram of the simulated backscattered fields from the tags

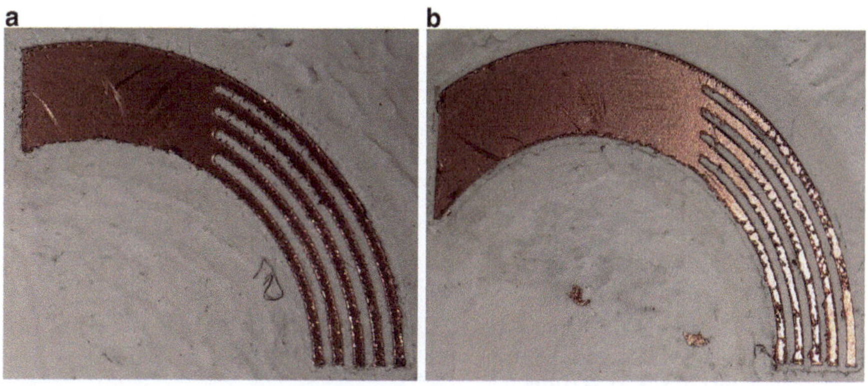

Fig. 3.32 Two 4-bit fabricated tags (**a**) $d_2 = 3$ mm and (**b**) $d_2 = 7$ mm

four resonant frequencies of the slots, five resonant frequencies are seen in the backscattered signal from the tags. In these circumstances, the resonant frequency of the metal shall be distinguished from the slot resonances in the detection process of the tag. The pole diagram of the tags is shown in Fig. 3.34 based on the measured backscattered signal. It is seen that the CNRs of the tags calculated from the measurement data are in good agreement with the simulation results.

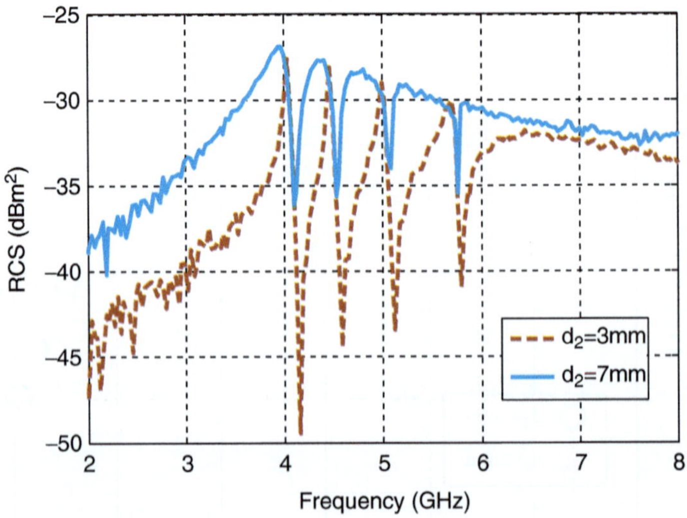

Fig. 3.33 Measured RCS of the tags

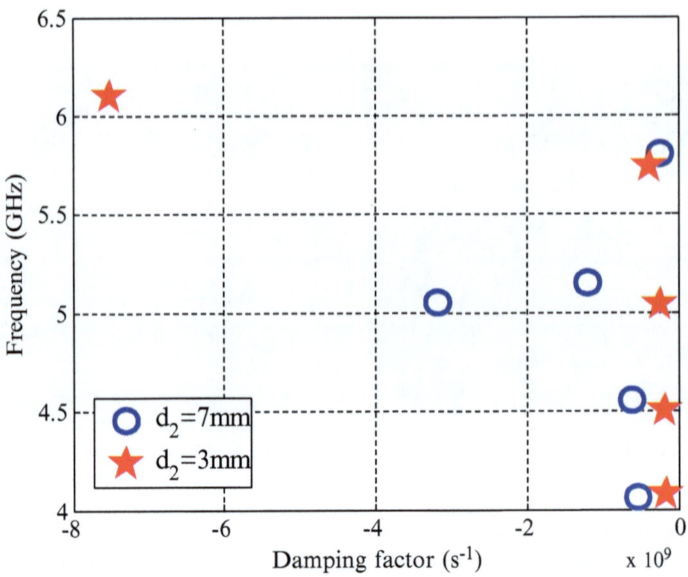

Fig. 3.34 Pole diagram of the measured backscattered fields from the tags

References

1. Preradovic S, Karmakar NC (2010) Chipless RFID: barcode of the future. IEEE Microw Mag 11(7):87–97
2. Hartmann CS (2002) A global SAW ID tag with large data capacity. IEEE Proceedings, ultrasonics symposium, pp 65–69
3. Blischak AT, Manteghi M (2011) Embedded Singularity Chipless RFID Tags. IEEE Trans Antennas Propag 59:3961–3968
4. Varahramyan CA (2006) Transmission delay line based ID generation circuit for RFID applications. IEEE Microw Wirel Compon Lett 16(11):588–590
5. Vemagiri J, Chamarti A, Agarwal M, Varahramayan K (2007) Transmission line delay-line radio frequency identification (RFID) tag. Microw Opt Technol Lett 49(8):1900–1904
6. Gupta S, Nikfal B, Caloz C (2011) Chipless RFID system based on group delay engineered dispersive delay structures. IEEE Antennas Propag Lett 10:1366–1368
7. Rezaiesarlak R, Manteghi M (2014) Complex-natural-resonance-based design of chipless RFID tag for high-density data. IEEE Trans Antennas Propag 62(2):898–904
8. Rezaiesarlak R, Manteghi M (2013) A low profile multi-bit chipless RFID tag. USNC-URSI meeting
9. Jalali I, Robertson ID (2005) Capacitively-tuned split microstrip resonators for RFID barcodes. European microwave conference
10. McVay J, Hoofer A, Engheta N (2006) Space-filling curve RFID tags. IEEE radio and wireless symposium, pp 199–202
11. Manteghi M, Rahmat-Samii Y (2007) Frequency notched UWB elliptical dipole tag with multi-bit data scattering properties. In Antennas and Propagation Society International Symposium, pp 789–792
12. Preradovic S, Balbin I, Karmakar NC, Swiegers GF (2009) Multiresonator-based chipless RFID system for low-cost item tracking. IEEE Trans Microw Theory Tech 57(5):1411–1419
13. Costa F, Genovesi S, Monorchio A (2013) A chipless RFID based on multiresonant high-impedance surfaces. IEEE Trans Microw Theory Tech 61(1):146–153
14. Rezaiesarlak R, Manteghi M (2014) On the application of characteristic modes for the design of chipless RFID tags. APS/URSI 2014, Memphis, TN
15. Vena A, Perret E, Tedjini S (2012) Design of compact and auto-compensated single-layer chipless RFID tag. IEEE Trans Microw Theory Tech 60:2913–2924
16. Baum CE (ed) (1997) Detection and identification of visually obscured targets. Taylor & Francis, Philadelphia, PA

Chapter 4
Identification of Chipless RFID Tags in the Reader

4.1 Introduction

The signal received by the antenna includes the scattered signal from the chipless RFID tag located in the reader area, reflections from background objects (clutter), and noise. Therefore, the detection process is a challenging aspect in the design of chipless RFID systems. In most situations, the reflections from background objects are stronger than the tag response. The identification can be performed based on the time-domain or frequency-domain response. In this chapter, first the time and frequency characteristics of the scattered signal from the tag are presented. It will be shown that the absolute value of the scattered field from the tag and the group delay response do not provide sufficient information about the resonant behavior of the tag. Then, a review on the time-frequency techniques, suitable for extracting the information from the signal, is presented. Two basic techniques such as short-time Fourier transform (STFT) and wavelet transform are studied in more detail. A new technique, called short-time matrix pencil method (STMPM) is introduced by which the resonant frequencies and damping factors of the CNRs are extracted from the time-domain response. It will be shown that by increasing the filtering parameter and decreasing the length of the window simultaneously, the resolution in both time and frequency domains can be improved, simultaneously. The performance of the method against noise and some important parameters such as resolution in time and frequency are studied in detail in this chapter. As an example, the time-frequency analysis of the open-ended cylinder is presented and compared to the results obtained from short-time frequency transform (STFT). Then, the application of the proposed technique in decoding the IDs of the chipless RFID tags is presented.

R. Rezaiesarlak, M. Manteghi, *Chipless RFID*, DOI 10.1007/978-3-319-10169-9_4

4.2 Scattered Signal from Chipless RFID Tags

As a general case, a bi-static set-up comprised of two UWB antennas is depicted in Fig. 4.1. The transmitter antenna, TX, is excited with an incident voltage $a_{TX}(t)$ and the radiated field from the transmitter at the scatterer position is $\mathbf{E}^{\text{inc}}(\mathbf{r}, t)$. The induced current on the surface of the scatterer re-radiates an electric field, $\mathbf{E}^{\text{s}}(\mathbf{r}_{\text{RX}}, t)$ at the position of the receiving antenna, RX. The scattered field generates a voltage $v_{\text{oc}}(t)$ across the receiving port. The transmission coefficient between two antennas is defined as the ratio of the received voltage wave $b_{RX}(s)$ to the incident voltage wave, $a_{TX}(s)$, in the Laplace domain. As the first step, we introduce the impulse response of a scatterer located at the origin as the $\hat{\mathbf{a}}_2$-component of the scattered field at $\mathbf{r}_2$, in the far zone of the scatterer, for an impulsive incident plane wave of $\hat{\mathbf{a}}_1\delta\left(t+\frac{\hat{\mathbf{r}}_1 . \mathbf{r}}{c}\right)$ as $\frac{\hat{\mathbf{a}}_1\hat{\mathbf{a}}_2}{r_2}\Gamma(\hat{\mathbf{r}}_1, \hat{\mathbf{r}}_2, t)$. In the general case, $\hat{\mathbf{a}}_1$ and $\hat{\mathbf{a}}_2$ are functions of time. The impulse response of the scatterer is a dyadic, which includes all the scatterer information for a particular $(\hat{\mathbf{r}}_1, \hat{\mathbf{r}}_2, t)$.

The next step is computing the incident field versus the effective length of the transmitting antenna and the input signal. The antenna can be characterized in the far-zone by its equivalent effective length [1, 2] as

$$\mathbf{h}(\hat{\mathbf{r}}; t) = -\frac{\hat{\mathbf{r}} \times \hat{\mathbf{r}}}{I_a} \times \int_{S_{ant}} \mathbf{J}\left(\mathbf{r}^{'}; t + \mathbf{r}^{'} \cdot \hat{\mathbf{r}}/c\right) dS^{'} \tag{4.1}$$

where S_{ant} indicates the antenna surface, $\mathbf{J}$ is the electric current on the antenna surface, $\hat{\mathbf{r}}$ is the unit vector to the observation point, I_a is the input current, and

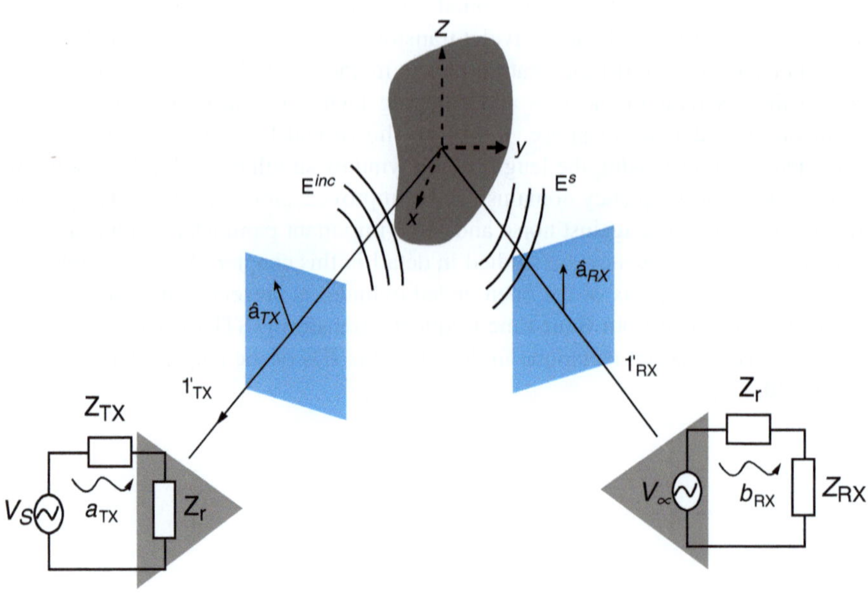

Fig. 4.1 Schematic of the bi-static set-up for measuring the impulse response of the tag

c represents the speed of light in free space. The primed and unprimed coordinates indicate the source and observation points, respectively. If the input impedance of the antenna is assumed as Z_r and Z_{TX} is the impedance of the matching circuit, then the reflection coefficient at the transmitter, S_{TX}, in the frequency domain is given by

$$S_{TX} = \frac{Z_{TX} - Z_r^*}{Z_{TX} + Z_r} \tag{4.2}$$

The realized effective length of the antenna in the time domain is defined as [2, 3]

$$\mathbf{h}_{TX}^{R}(\hat{\mathbf{r}}; t) = \frac{1 - s_{TX}(t)}{2} * \mathbf{h}_{TX}(\hat{\mathbf{r}}; t) \tag{4.3}$$

where $s_{TX}(t)$ is the inverse Fourier transform of S_{TX}. Using the definition of the realized effective length of the transmitting antenna, the incident electric field $\mathbf{E}^{inc}$ for an incident voltage wave of $a_{TX}(t)$ is computed from

$$\mathbf{E}^{inc}(\mathbf{r}_{TX}; t) = -\mu \frac{1}{2\pi r_{TX}} \frac{\partial}{\partial t}\left[\frac{a_{TX}(t)}{\sqrt{R_{TX}}} * \mathbf{h}_{TX}^{R}\left(-\hat{\mathbf{r}}_{TX}; t - \frac{r_{TX}}{c}\right)\right] \tag{4.4}$$

where R_{TX} is the resistance of the transmitter and μ is the permeability of the free space. Using the definition of the impulse response of the scatterer, one can compute the received signal at the receiving antenna port as

$$\begin{aligned} b_{RX}(t) = \frac{-\mu}{8\pi^2 r_{TX} r_{RX}} \frac{\partial}{\partial t}\Bigg[\frac{a_{TX}(t)}{\sqrt{R_{RX}}\sqrt{R_{TX}}} * \mathbf{h}_{TX}^{R}\left(-\hat{\mathbf{r}}_{TX}; t - \frac{r_{TX}}{c}\right) \\ * \hat{\mathbf{a}}_1 \hat{\mathbf{a}}_2 \Gamma\left(\hat{\mathbf{r}}_{TX}, \hat{\mathbf{r}}_{RX}; t - \frac{r_{RX}}{c}\right) * \mathbf{h}_{RX}^{R}(-\hat{\mathbf{r}}_{RX}; t)\Bigg] \end{aligned} \tag{4.5}$$

Equation (4.5) can be simplified if the impulse response of the scatterer is computed for the case where $\hat{\mathbf{a}}_1$ and $\hat{\mathbf{a}}_2$ are in the same direction as $\mathbf{h}_{TX}^{R}$ and $\mathbf{h}_{RX}^{R}$, respectively.

$$\begin{aligned} b_{RX}(t) = \frac{-\mu}{8\pi^2 r_{TX} r_{RX}} \frac{\partial}{\partial t}\Bigg[\frac{a_{TX}(t)}{\sqrt{R_{RX}}\sqrt{R_{TX}}} * \mathbf{h}_{TX}^{R}\left(-\hat{\mathbf{r}}_{TX}; t - \frac{r_{TX}}{c}\right) \\ * \boldsymbol{\Gamma}\left(\hat{\mathbf{r}}_{TX}, \hat{\mathbf{r}}_{RX}; t - \frac{r_{RX}}{c}\right) * \mathbf{h}_{RX}^{R}(-\hat{\mathbf{r}}_{RX}; t)\Bigg] \end{aligned} \tag{4.6}$$

Applying Laplace transform to (4.6), the transmission coefficient is defined as the ratio of b_{RX}(s) to a_{TX}(s).

$$S_{RT}(s) = \frac{s\mu e^{-s(r_{TX}+r_{RX})/c}}{8\pi^2 r r'' \sqrt{R_{TX} R_{RX}}} \mathbf{H}_{TX}^{R}(-\hat{\mathbf{r}}_{TX}; s) \cdot \Gamma(\hat{\mathbf{r}}_{TX}, \hat{\mathbf{r}}_{RX}; s) \cdot \mathbf{H}_{RX}^{R}(\hat{\mathbf{r}}_{RX}; s) \tag{4.7}$$

For the mono-static case, where one antenna is used for both transmitting and receiving, the reflection coefficient is computed from

$$S_{11} = \frac{s\mu e^{-2sr_{RX}/c}}{8\pi^2 {r_{RX}}^2 R_{RX}} \left|\mathbf{H}_{TR}^{R}(\hat{\mathbf{r}}; s)\right|^2 \Gamma(, \hat{\mathbf{r}}, \hat{\mathbf{r}}; s) \tag{4.8}$$

From (4.8), the impulse response of the scatterer is obtained as

$$\Gamma(, \hat{\mathbf{r}}, \hat{\mathbf{r}}; s) = \frac{8\pi^2 {r_{RX}}^2 R_{TR} e^{2sr_{RX}/c}}{s\mu \left|\mathbf{H}_{TR}^{R}(\hat{\mathbf{r}}; s)\right|^2} S_{11} \tag{4.9}$$

According to (4.9), all the structural information of the scatterer is included in the reflection coefficient of the receiving antenna normalized by the square of the magnitude of the effective length of the antenna.

Based on the singularity expansion method (SEM), the impulse response of the scatterer for the incident and scattered fields directed in the $\hat{\mathbf{r}}_1$ and $\hat{\mathbf{r}}_2$ can be expressed by

$$\Gamma(\hat{\mathbf{r}}_1, \hat{\mathbf{r}}_2; s) = \sum_{n=1}^{N} \frac{R_n(\hat{\mathbf{r}}_1, \hat{\mathbf{r}}_2)}{s - s_n} + \Gamma_e(\hat{\mathbf{r}}_1, \hat{\mathbf{r}}_2; s) \tag{4.10}$$

where the first term is the late-time response, including the complex natural resonances $s_{\rm n}$ and corresponding residues $R_{\rm n}$, and the second term is the early-time response of the scatterer. Compared to the early-time response and residues, the complex natural resonances (CNRs) are aspect-independent. Although each scatterer includes an infinite number of CNRs, the series in (4.10) is truncated to N, the number of fundamental resonances excited by the incident electric field. Compared to the CNRs, the residues $R_{\rm n}$ and the early-time response of the tag $\Gamma_{\rm e}$ depend strongly on the direction and polarization of the transmitting and receiving antennas. As mentioned in Chap. 2, the embedded CNRs of the tag are high-Q resonances. Assuming the tag is illuminated by an incident pulse $\delta(t)$, the scattered field in close proximity of the nth resonant frequency is written by

$$E(\omega) = \frac{R_n}{(\alpha - \alpha_{\rm n}) + j(\omega - \omega_{\rm n})} + \sum_{\substack{m \neq n \\ m = 1}}^{N} \frac{R_{\rm m}}{(\alpha - \alpha_{\rm m}) + j(\omega - \omega_{\rm m})} + \Gamma_{\rm e}(\omega) \tag{4.11}$$

Although the late-time response has its maximums at the resonant frequencies of the tag, it does not necessarily happen for the total field. The received signal at the nth resonant frequency is

$$E(\omega_{\rm n}) = \frac{R_{\rm n}}{(\alpha - \alpha_{\rm n})} + \sum_{\substack{m \neq n \\ m = 1}}^{N} \frac{R_{\rm m}}{(\alpha - \alpha_{\rm m}) + j(\omega_{\rm n} - \omega_{\rm m})} + \Gamma_{\rm e}(\omega_{\rm n}) \tag{4.12}$$

As (4.12) shows, the received signal at $\omega = \omega_{\rm n}$ is separated into three terms. The first term in (4.12) is the scattered field from the nth resonant frequency; the second term is due to the coupling of the other resonators on the nth resonator resonating at $\omega = \omega_{\mathrm{n}}$, and the third term is the early-time response of the tag at ω_{n}. As an important note here, the magnitude of the scattered field at the nth resonant frequency is not simply a maximum at ω_{n}. The coupling of the other poles, the second term in (4.12), and early-time response can change the maximum peak of the total field in the frequency domain to a minimum null or may shift it to other frequencies. Since the couplings and the early-time response of the tag are aspect-dependent, the magnitude of the scattered field in the frequency domain is aspect-dependent as well. The main observation is that the magnitude of the impulse response of the tag in the frequency domain is not sufficient to extract the resonant frequencies of the tag.

Similarly, in detection techniques based on the group delay of the received signal, the group delay is not sufficient for extracting the resonant frequencies of the tag. The group delay is defined as

$$\tau_{\mathrm{g}} = -\frac{d\phi_{s_{11}}(\omega)}{d\omega} \tag{4.13}$$

where $\varphi_{s_{11}}(\omega)$ is the phase response of the tag. Again, assuming high-Q resonances are embedded on the tag and ignoring the effect of the second term in (4.12), the group delay of the received signal can be written by

$$\tau_{\mathrm{g}} = \frac{1}{1 + \left(\frac{\omega - \omega_{\mathrm{n}}}{\alpha - \alpha_{\mathrm{n}}}\right)^2} - \frac{d\phi_{\Gamma_{\mathrm{e}}}(\omega)}{d\omega} \tag{4.14}$$

where $\phi_{\Gamma_e}(\omega)$ is the phase of the early-time part of the signal, which is aspect-dependent. Therefore, both the magnitude and phase of the impulse response of the scatterer can vary by changing the source and observation points. Ignoring the early-time part of the signal causes a misleading result where the group delay shows its maximum value at the resonant frequencies of the tag. In the cases where the phase of the early-time part has stronger variations than the phase of the late-time part, the detection of the resonant frequencies from the group delay is not straightforward. As an example, Fig. 4.2 shows a single-bit tag, located in the xy plane, illuminated by an incident electric field. The scattered signal in given two different orientations of the receiving antenna is depicted in Fig. 4.3. The dominantly excited resonant frequency of the tag is $f = 5.09$ GHz, while the peaks and nulls in the scattered signal are slightly shifted around it. Hence, the locations of nulls and peaks in the backscattered response from the tag are not the exact values of the resonant frequencies of the tag. This is very important in the identification of a tag with high density of data, in which case the resonant frequencies are close to each other. In Fig. 4.4, the group delay of the received signal for two different cases is shown. Although the first peak of the group delay is exactly located at the resonant frequency of the slot, there are some other peaks corresponding to variations in the

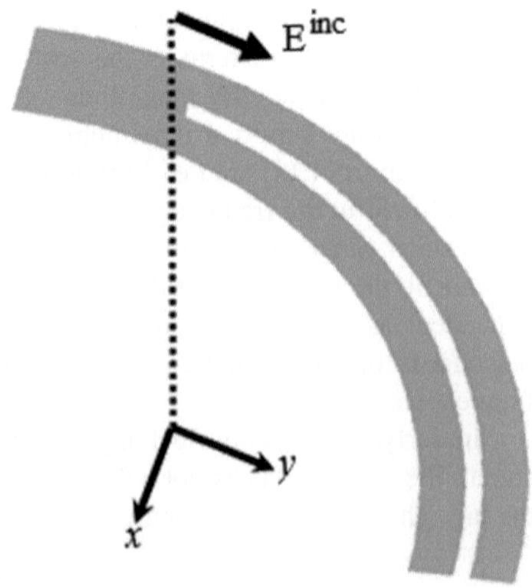

Fig. 4.2 Single-bit tag illuminated by a plane incident field

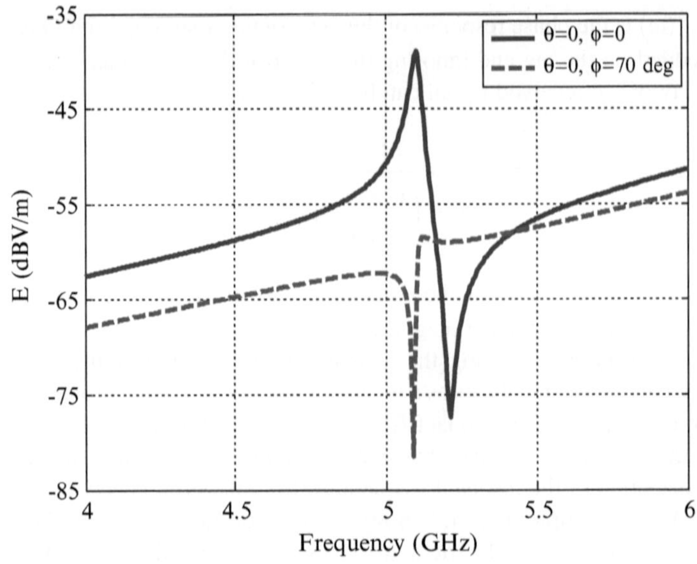

Fig. 4.3 Scattered electric field from the tag for two different orientations of receiving antenna

early-time response. Hence, the absolute value of the scattered signal and its group delay cannot be used to accurately extract the ID of the tag.

In contrast to the absolute-value and group-delay response of the scattered field, the detection can be performed based on the time-domain response. As mentioned before, the time-domain response of the tag is the combination of the early-time and late-time responses. The aspect-independent parameters of the tag, the complex natural resonances (CNRs), are included in the late-time response. This part of the

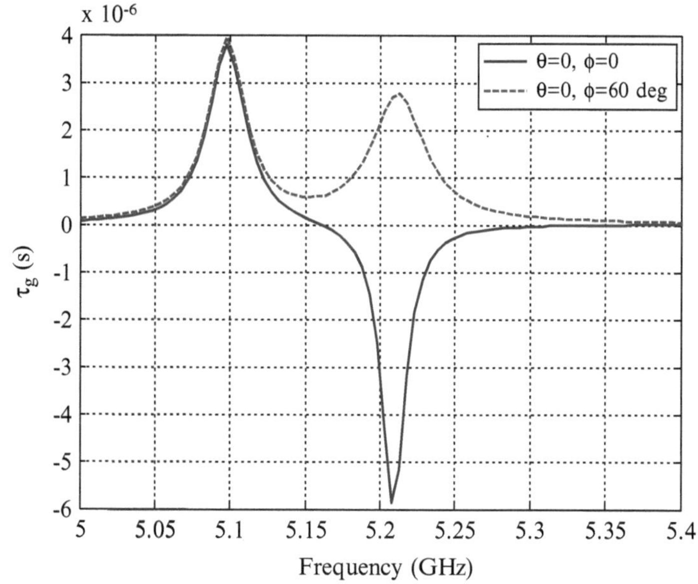

Fig. 4.4 Group delay of the scattered field for two different orientations of receiving antenna

response must be separated from the early time, which contains the specular reflections from the tag. Therefore, we need to employ an efficient time-frequency approach by which one can extract the turn-on times of the resonances and the CNRs.

4.3 Time-Frequency Analysis of the Scattered Signal

The basic idea of the Fourier transform is that any arbitrary signal in the time domain can be decomposed into a set of sinusoidal signals of different frequencies, phases, and magnitudes. Using the orthogonality property of sinusoidal functions, the magnitudes and phases (complex magnitudes) are found to be functions of frequency. There are two issues with the Fourier transform of a scattered signal, which may limit the application of a pure Fourier transformation for detection applications. The first issue arises when different components of the scattered signal are received at different instants in time. The classical Fourier transform uses the entirety of the time-domain signal to find its frequency response. Therefore, the information about the time of arrival will be lost in the Fourier transformation. Additionally, in practical applications, the loss mechanisms in the scattering medium attenuate the amplitude of the scattered signal. Therefore, instead of working with unlimited time-domain sinusoidal signals, we are dealing with damped sinusoidal signals with different turn-on times. In the time-frequency approaches, the local frequency contents of the non-stationary signal are represented versus time.

4.3.1 Short-Time Fourier Transform

The simplest time-frequency approach is the short-time Fourier transform (STFT). In this technique, a sliding window moves along the time axis and the frequency contents of the signal are represented at each snapshot of the time. In other words, the non-stationary signal is divided into a sequence of quasi-stationary signals. The frequency components of each time segment are obtained by applying Fourier transform to the windowed signal. The STFT of a time-domain signal $x(t)$ is a function of time τ and frequency ω and is defined as [4]

$$X(\tau,\omega) = \int_{-\infty}^{+\infty} x(t)h(t-\tau)e^{-j\omega t}dt \tag{4.15}$$

where $h(t)$ is the window function, usually a Gaussian window centered around zero, and $x(t)$ is the time-domain signal. The windowed signal is projected onto a set of basis functions $h(t-\tau)e^{-j\omega t}$. Since the basis functions are not infinite in time, the local frequency components of the signal can be monitored in terms of time. When τ and ω are continuous functions, the time-domain signal can be reconstructed from the inverse short-time Fourier transform as

$$x(t) = \frac{1}{2\pi}\iint X(\tau,\omega)h(t-\tau)e^{j\omega t}d\omega \cdot d\tau \tag{4.16}$$

which holds for any arbitrary chosen window $h(t)$. An important parameter in STFT is the length of the window, which affects resolution in the time and frequency domains.

The width Δt of the sliding window $h(t)$ in the time domain and its corresponding bandwidth $\Delta\omega$ in the frequency domain are defined as [5]

$$\Delta t^2 = \frac{\int t^2|h(t)|^2dt}{\int |h(t)|^2dt} \tag{4.17}$$

$$\Delta\omega^2 = \frac{\int \omega^2|H(\omega)|^2d\omega}{\int |H(\omega)|^2d\omega} \tag{4.18}$$

where $H(\omega)$ is the Fourier transform of $h(t)$ and the denominator is the energy of the window. Since the two pulses can only be distinguished in the time domain if they are more than Δt apart, Δt represents the time resolution of the window. Similarly, two sinusoidal signals can be distinguished only if they are more than $\Delta\omega$ apart [5]. Hence, Δt and $\Delta\omega$ define the time and frequency resolutions, respectively. Based on the uncertainty principle, the product of the resolutions in time and frequency domains is limited by

$$\Delta\omega \cdot \Delta t \geq \frac{1}{2} \tag{4.19}$$

According to (4.19), more improvement in the time resolution out of certainty principle, will deteriorate the frequency resolution and vice versa. The most common window function used in STFT is the Gaussian function, defined as

$$h(t) = \frac{1}{\sqrt{2\pi}\delta} \exp\left(-\frac{t^2}{2\delta^2}\right) \tag{4.20}$$

The Fourier transform of h(t) is also a Gaussian function

$$H(\omega) = \exp\left(-\frac{\delta^2\omega^2}{2}\right) \tag{4.21}$$

The quantity δ is the variance of the Gaussian function. From (4.17) and (4.18), the resolution in time and frequency for the Gaussian function is given by

$$\Delta t^2 = \frac{\delta^2}{2} \tag{4.22}$$

$$\Delta\omega^2 = \frac{1}{2\delta^2} \tag{4.23}$$

which satisfies the minimum time-bandwidth product $\Delta t \cdot \Delta\omega = 1/2$. The performance of STFT is strongly dependent upon the window function.

As an example, consider the following time-domain signal

$$s_1(t) = \begin{cases} 3e^{-\alpha_1 t} \sin(2\pi f_1 t) & 0 \leq t \leq t_0 \\ 3e^{-\alpha_1 t} \sin(2\pi f_1 t) + 2e^{-\alpha_2 t} \sin(2\pi f_2 t) & t_0 < t \end{cases} \tag{4.24}$$

The signal in (4.24) contains two sinusoidal signals with turn-on times at $t = 0$ and $t = t_0$. It emulates the backscattered signal from two tags with resonant frequencies spaced apart from each other. The first resonant frequency at $f = f_1$ is corresponding to the first tag, and then the second tag's resonant frequency starts at $t = t_0$. For simplicity in discussion, the early-time responses of the tags are not included in the analysis. The time-domain response in (4.24) is depicted in Fig. 4.5 for $f_1 = 5$GHz, $f_2 = 6$ GHz, $\alpha_1 = 7$e8, $\alpha_2 = 1$e9, and $t_0 = 5$ ns. By applying STFT to the time-domain response, the spectrograms of the signal are shown in Fig. 4.6 for

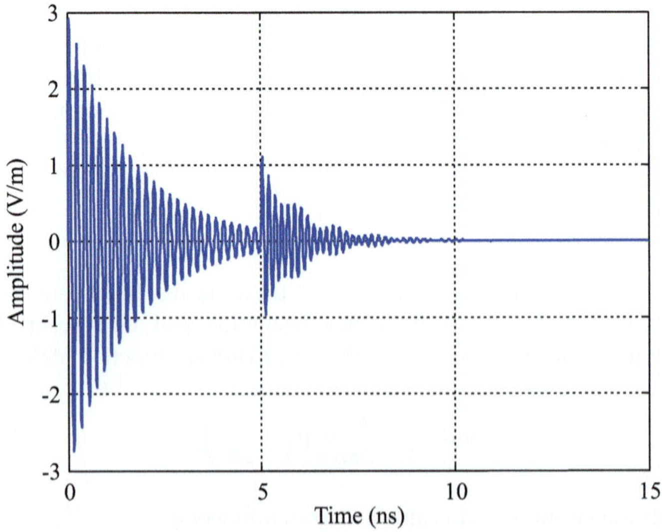

Fig. 4.5 Time-domain signal emulating scattered signal from two single-bit tags

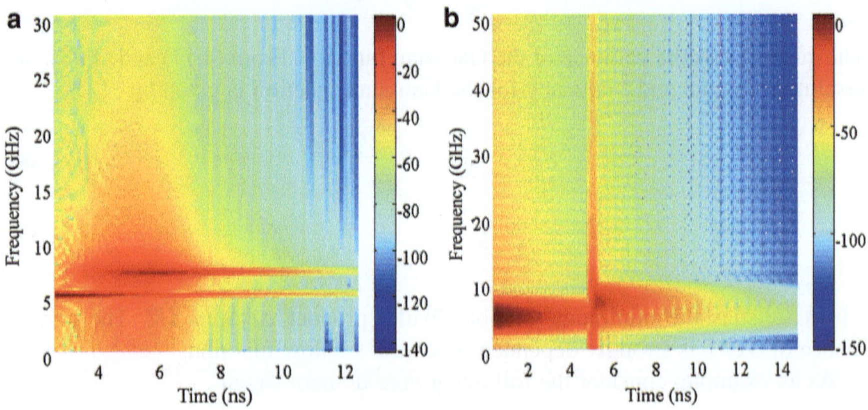

Fig. 4.6 Spectrogram of the signal for two window lengths of (**a**) 5 ns (**b**) 0.64 ns

two different window lengths of $T = 5$ ns and 0.64 ns. According to Fig. 4.6a, by increasing the window's length and thereby increasing the frequency resolution, the two resonant frequencies of the signal are easily distinguished. Still, the turn-on time of the second tag cannot be clearly seen in the spectrogram. In order to improve the time resolution, the window length is decreased to $T = 0.64$ ns in Fig. 4.6b. In this case, the turn-on time is clearly visible, but the frequency resolution deteriorates as a result of the small window length.

For high density of data, in which we need to pack the resonant frequencies of the tag into a narrow frequency band, STFT is not suitable to distinguish the resonant frequencies of the tags and simultaneously obtain the turn-on times of the resonances. This limitation is more pronounced when multiple tags are present in the reader zone.

4.3.2 Wavelet Transform

The main drawback of STFT is that it has a fixed resolution in time and frequency domains. Because of the non-stationary behavior of the backscattered signals, we need to change the resolution of the window in the time and frequency domains in order to extract the required information from the signal.

The location and ID of the tag can be extracted from the time-frequency analysis of the backscattered signal. A better resolution in time domain improves the accuracy of the calculated turn-on time and better resolution in the frequency domain enhances the identification process of the tags. For these applications, STFT is not a suitable approach.

Wavelet transform is another approach, which can be used for the time-frequency analysis of the backscattered signals from a tag. Compared to STFT, the wavelet transform is a time-frequency representation of the signal, which achieves the variable resolution in one domain and multi-resolution in the other domain. The continuous wavelet transform (CWT) of signal $x(t)$ is defined as

$$W_{\mathrm{h}}^{\mathrm{x}}(s,\tau) = \langle x, h_{\mathrm{s},\tau} \rangle = \int_{R} x(t) h_{\mathrm{s},\tau}(t) dt \tag{4.25}$$

where R is the integration limit and $h_{s,\tau}$ is the wavelet function defined by [4, 5]

$$h_{\mathrm{s},\tau}(t) = \frac{1}{\sqrt{s}} h\left(\frac{t-\tau}{s}\right) \tag{4.26}$$

where s is a positive scaling factor and τ is a real number which shifts the child wavelet h. The wavelets are dilated when the scale $s > 1$ and are contracted when $s < 1$. The constant $s^{-1/2}$ in (4.26) is for energy normalization. Therefore

$$\int |h_{\mathrm{s},\tau}|^2 dt = 1 \tag{4.27}$$

The wavelet coefficients of the signal are represented in the scalogram. Different types of wavelets have been proposed for different applications. In Fig. 4.7, some of the more well-known wavelets are shown. Depending on the shape of the signal for a specific application, a particular wavelet may be more appropriate. As an example, for the backscattered signal from resonant-based structures, the Morlet wavelet is more efficient because of the similarity of the wavelet to the sinusoidal form of the signal in the late time. The scalogram of the signal seen in (4.24) is represented

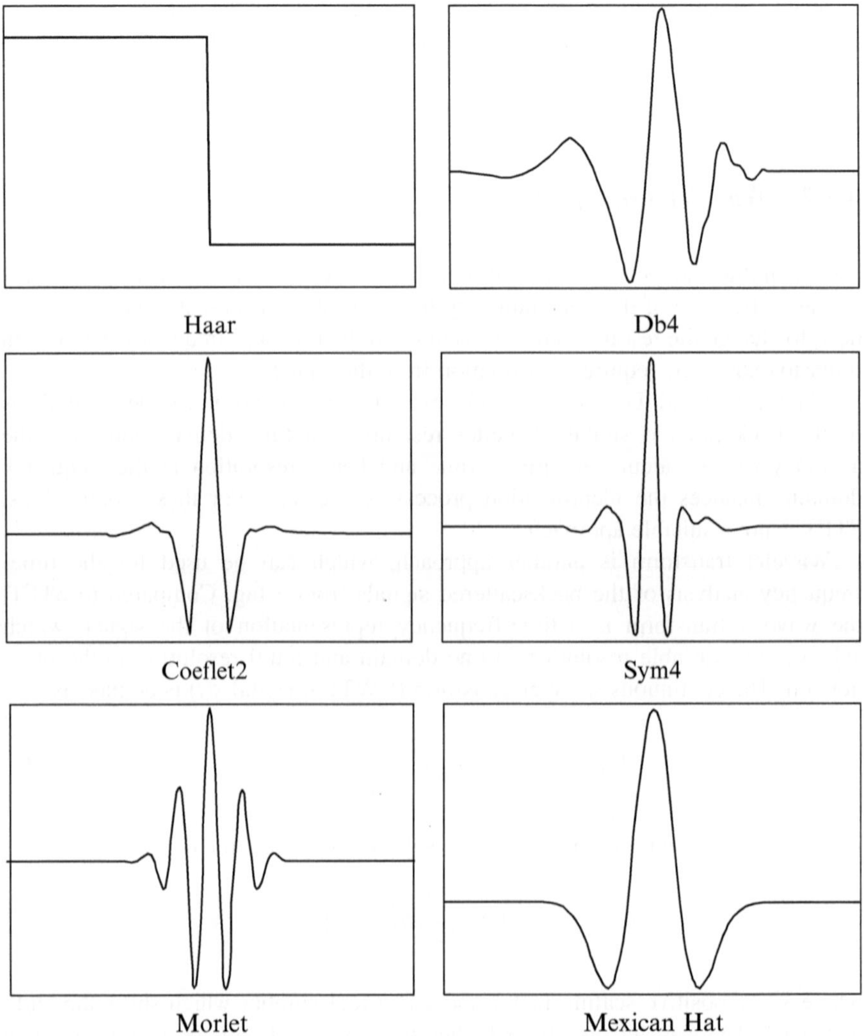

Fig. 4.7 Some practical wavelets

in Fig. 4.8. The turn-on times and the resonant frequencies of the signal can be obtained from its time-frequency diagram. The wavelet coefficients shown in Fig. 4.8 provide significant information of the tag. Because of the existence of high-frequency components near the turn-on times of the tags, the wavelet coefficients obtained from sliding the wavelet with low scale factor along the time axis gives us the turn-on times of the tags. As an example, Fig. 4.9 shows the wavelet coefficients of the scalogram by applying the wavelet of $s = 2$ along the time axis.

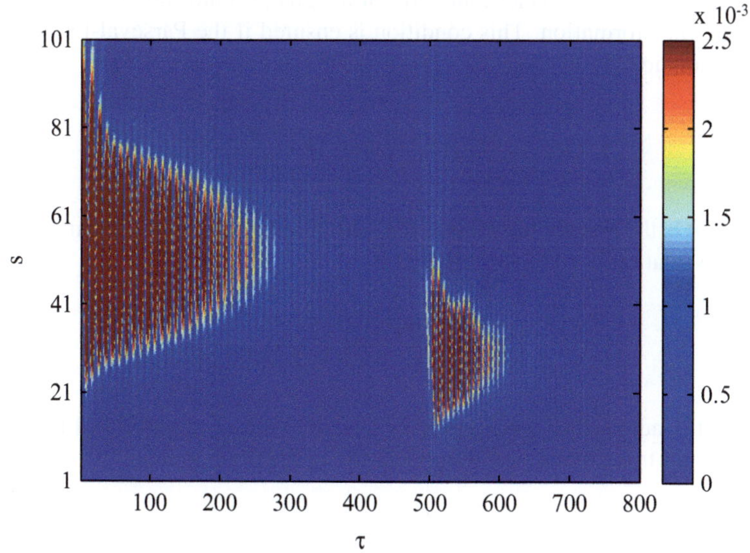

Fig. 4.8 Scalogram of the signal with Morlet wavelet

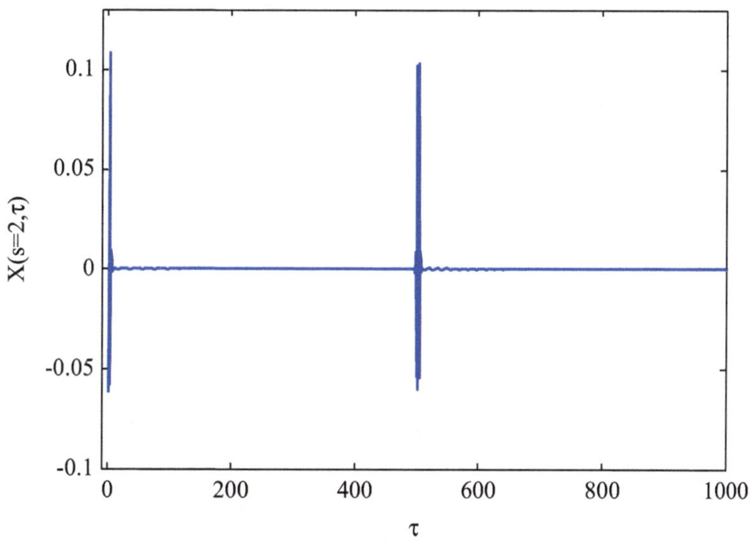

Fig. 4.9 Wavelet coefficient versus τ by applying Morlet wavelet with $s = 2$

For a specific application, one may generate one's own wavelet function. Based on wavelet theory, each wavelet function should satisfy some important properties in order to be efficiently employed in time-frequency analysis of the signal. The most important properties of the wavelet are the admissibility and regulatory conditions [5]. In the wavelet transform of a signal, no information should be lost during the transformation. This condition is ensured if the Parseval-like condition (4.28) is satisfied [5].

$$\int\int \langle f_1, h_{s,\tau}\rangle \langle h_{s,\tau}, f_2\rangle \frac{ds}{s^2} d\tau = \int \frac{|H(\omega)^2|}{|\omega|} \langle f_1, f_2\rangle d\omega \tag{4.28}$$

where $H(\omega)$ is the Fourier transform of $h(t)$ and f_1 and f_2 are two arbitrary functions. The above relation in (4.28) is satisfied if

$$\int \frac{|H(\omega)^2|}{|\omega|} d\omega < +\infty \tag{4.29}$$

Based on the admissible condition in (4.29), the Fourier transform of the wavelet must be equal to zero at $\omega = 0$. Equivalently, in the time domain

$$\int h(t)dt = 0 \tag{4.30}$$

Hence, the wavelet function should have an oscillatory behavior in the time domain.

The other important property of the wavelet is the regulation condition. According to this condition, the wavelet should be local in both time and frequency domains. Aforementioned conditions introduce the wavelet as a tool for multiresolution signal analysis with the varying scale factor, s. This means that the original signal can be decomposed into multiple frequency bands and each component can be analyzed independently, which is very useful in analyzing the scattered signal from chipless RFID tags. In each band, the signal is analyzed with a resolution matched to the scale of the wavelets. Assuming the scale of the wavelet is s, then the corresponding bandwidth of the wavelet is [5]

$$(\Delta W_s)^2 = \frac{\int \omega^2 |H(s\omega)^2| d\omega}{\int |H(s\omega)^2| d\omega} = \frac{1}{s^2}(\Delta W)^2 \tag{4.31}$$

where the last equality is obtained from (4.23). The fidelity factor of the wavelet is defined by the central frequency divided by the bandwidth of the wavelet. Hence, the fidelity factor is given by

$$\begin{aligned}\frac{1}{Q} &= \frac{\Delta W_s}{1/s} \\ &= \Delta W\end{aligned} \tag{4.32}$$

As (4.32) shows, the relative bandwidth of the wavelet does not change with the scale factor. It means that for high frequencies, the wavelet has larger bandwidth (corresponding to low scale factor s) which leads to a narrow time window in the time domain and vice versa. This adaptive property of the wavelet is very useful in the time-frequency analysis of the scattered signals from the tag.

The main drawback of the wavelet in chipless RFID applications is its low frequency resolution for high density of data. Several time-frequency representations such as Wigner-Vile distribution, adaptive time-frequency techniques, bi-linear time-frequency transform, and so on [6] have been proposed for this application, which introduce some pros and cons. By improving the resolution in each technique, some interferences are presented in the time-frequency diagram of the signal which may provide some difficulties in the identification process of the signal. Common in all techniques is the continuous form of the result in the time-frequency diagram. In chipless RFID applications, the ID of the tag is the discrete resonant frequencies embedded on the tag. Hence, it is more convenient to illustrate the CNRs discretely in the time-frequency diagram of the signal. In the next section, short-time matrix pencil method is introduced as a time-frequency technique by which the turn-on time and resonant frequencies of the tags are shown discretely in the time-frequency diagram.

4.4 Short-Time Matrix Pencil Method (STMPM)

In the previous sections, the STFT and wavelet transform were studied for the detection of chipless RFID tags. Both techniques have difficulties in extracting the CNRs in applications where a high density of data is encoded on the tag. In this section, a time-frequency technique based on matrix pencil method is presented. As previously mentioned, the late-time response can be written as the summation of damped sinusoidals with weighting residues. Since the ID of the tag is included in the CNRs of the response, an accurate detection technique for extracting the CNRs is critical. Various techniques such as Prony's method [7], matrix pencil method (MPM) [8], E-pulse [9], Cauchy method [10], and more [11] have been proposed for extracting the resonant frequencies of the scattered signal from resonant-based structures. Among the proposed techniques, MPM, employing singular value decomposition (SVD) shows the best performance against noise. Compared to the resonant frequencies of the CNRs, the damping factors are very sensitive to noise. The accuracy of the calculated CNRs can be improved by employing the proposed technique, called short-time matrix pencil method (STMPM). Since the proposed time-frequency approach is strongly based on the matrix-pencil method, we first present the idea of MPM in extracting the poles of a signal.

4.4.1 Matrix-Pencil Method

If two functions $g(t)$ and $h(t)$ are defined on a common interval, then

$$f(t, \lambda) = g(t) + \lambda h(t) \tag{4.33}$$

is called the pencil of the two functions with a scalar parameter λ [8]. Employing the pencil method, the transient response of the tag can be modeled as

$$y(t_k) = \sum_{n=1}^{N} R_n {z_n}^k \tag{4.34}$$

where

$$z_n = e^{s_n \Delta t} \tag{4.35}$$

in which N is the number of the poles in the signal and $k = 0,1, \ldots., N - 1$ is the sample index. R_n is the residue of the nth pole, s_n, and Δt is the sampling interval. In absence of noise, one can calculate $[Y_1]$ and $[Y_2]$ matrices using the sampled data.

$$[Y_1] = \begin{bmatrix} y(0) & y(1) & \ldots & y(L-1) \\ y(1) & y(2) & \ldots & y(L) \\ \vdots & \vdots & & \vdots \\ y(N-L-1) & y(N-L) & \ldots & Y(N-2) \end{bmatrix}_{(N-L)\times L} \tag{4.36}$$

$$[Y_2] = \begin{bmatrix} y(1) & y(2) & \ldots & y(L) \\ y(2) & y(3) & \ldots & y(L+1) \\ \vdots & \vdots & & \vdots \\ y(N-L) & y(N-L+1) & \ldots & Y(N-1) \end{bmatrix}_{(N-L)\times L} \tag{4.37}$$

It is easy to show that the matrices in (4.36) and (4.37) can be decomposed for an arbitrary $L < N - 1$ in the following manner

$$[Y_1] = [Z_1][R][Z_2] \tag{4.38}$$

$$[Y_2] = [Z_1][R][Z_0][Z_2] \tag{4.39}$$

where $[Z_0]$ and $[R]$ are diagonal matrices

$$[Z_0] = \text{diag}[z_1, z_2, \ldots, z_\text{N}] \tag{4.40}$$

$$[R] = \text{diag}[R_1, R_2, \ldots, R_\text{N}] \tag{4.41}$$

and

$$[Z_1] = \begin{bmatrix} 1 & 1 & \cdots & 1 \\ z_1 & z_2 & \cdots & z_M \\ \vdots & \vdots & & \vdots \\ {z_1}^{(N-L-1)} & {z_2}^{(N-L-1)} & \cdots & {z_N}^{(N-L-1)} \end{bmatrix}_{(N-L)\times N} \tag{4.42}$$

$$[Z_2] = \begin{bmatrix} 1 & z_1 & \cdots & {z_1}^{L-1} \\ 1 & z_2 & \cdots & {z_2}^{L-1} \\ \vdots & \vdots & & \vdots \\ 1 & z_\text{M} & \cdots & {z_\text{M}}^{L-1} \end{bmatrix}_{N\times L} \tag{4.43}$$

Now consider the following matrix pencil

$$[Y_2] - \lambda[Y_1] = [Z_1][R]\{[Z_0] - \lambda[I]\}[Z_2] \tag{4.44}$$

where [I] is an $N \times N$ identity matrix. It is easy to show that if $\lambda = z_\text{i}$ $(i = 1, \ldots, N)$, the ith row of $\{[Z_0] - \lambda[I]\}$ is zero. Therefore, z_i is the generalized eigenvalue of the matrix pair $\{[Y_2], [Y_1]\}$. Thus, the problem of solving for z_i is in solving the following ordinary eigenvalue problem.

$$\left\{[Y_1]^+[Y_2] - \lambda[I]\right\} \tag{4.45}$$

where $[Y_1]^+$ is the Moore-Penrose pseudo-inverse of $[Y_1]$.

For noisy data, some prefiltering needs to be applied in order to avoid extra poles resulting from noise. In this case, one forms the data matrix [Y] from the noisy signal as

$$[Y] = \begin{bmatrix} y(0) & y(1) & \cdots & y(L) \\ y(1) & y(2) & \cdots & y(L+1) \\ \vdots & \vdots & & \vdots \\ y(N-L-1) & y(N-L) & \cdots & y(N-L) \end{bmatrix}_{(N-L)\times(L+1)} \tag{4.46}$$

As can be seen, $[Y_1]$ and $[Y_2]$ are obtained from [Y] by removing the last and the first columns, respectively. For efficient noise filtering, parameter L is chosen to be between $N/3$ and $N/2$. A singular-value decomposition (SVD) of the matrix [Y] can be written as

$$[Y] = [U][\Sigma][V]^H \tag{4.47}$$

Where [U] and [V] are unitary matrices. The parameter M is introduced as a threshold and the singular values beyond M are set to zero. An appropriate way to choose M is through looking at the ratio of the maximum singular value to all other singular values in the matrix. Considering σ_c as the singular value such that

$$\frac{\sigma_c}{\sigma_{\max}} \approx 10^{-p} \tag{4.48}$$

where p is the number of significant decimal digits in the data. For instance, if the measurement data is known to be accurate up to two significant digits, then the singular values for which the ratio in (4.48) are less than 10^{-2} are the singular values of measured noise [8]. The next step is to construct the filtered matrix $[V']$. It is constructed such that it contains only M dominant right-singular vectors $[V]$.

$$\left[V'\right] = [v_1, v_2, \ldots, v_M] \tag{4.49}$$

Now, by suppressing the singularities corresponding to the noise, $[Y_1]$ and $[Y_2]$ can be constructed.

$$[Y_1] = [U]\left[\Sigma'\right]\left[V'_1\right]^H \tag{4.50}$$

$$[Y_2] = [U]\left[\Sigma'\right]\left[V'_2\right]^H \tag{4.51}$$

where $[V'_1]$ and $[V'_2]$ are obtained from $[V]$ by removing the first and last rows and $[\Sigma']$ is obtained from the Mth column of $[\Sigma]$, corresponding to the M dominant singular values. Following the same approach as in (4.44), the poles of the signal can be obtained. Once the poles and M are known, the residues R_i are found from the following least squares problem.

$$\begin{bmatrix} y(0) \\ y(1) \\ \vdots \\ y(N-1) \end{bmatrix} = \begin{bmatrix} 1 & 1 & \cdots & 1 \\ z_1 & z_2 & \cdots & z_M \\ \vdots & \vdots & & \vdots \\ z_1^{N-1} & z_2^{N-1} & \cdots & z_M^{N-1} \end{bmatrix} \begin{bmatrix} R_1 \\ R_2 \\ \vdots \\ R_M \end{bmatrix} \tag{4.52}$$

4.4.2 STMPM

In summary, the poles of the late-time response of the tag can be obtained by applying the matrix-pencil method to the time-domain data. An important parameter is the turn-on time of the poles at which the illuminating plane wave passes through the tag. In other words, we need a time-frequency approach in order to obtain the localized complex natural resonances of response. This will be useful in applications where multiple tags are present in the reader area. In this approach, a sliding time-window is introduced and the matrix-pencil method (MPM) is applied to extract the local CNRs and residues of the poles at each snapshot in time.

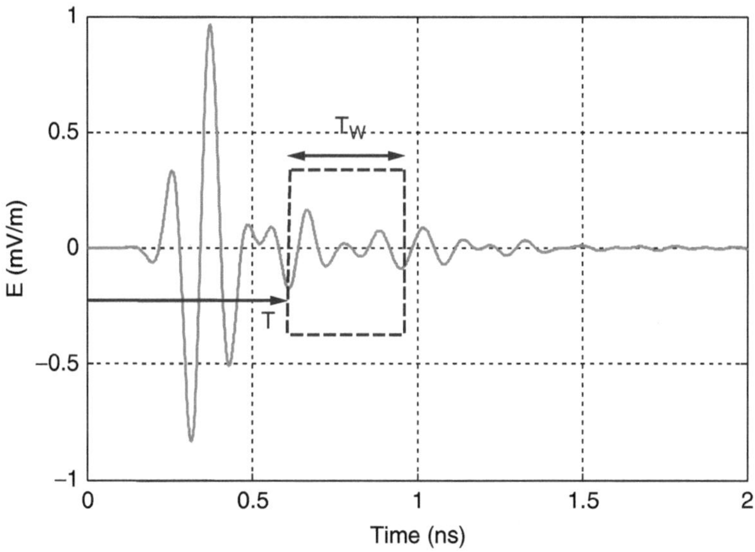

Fig. 4.10 Application of STMPM to the scattered signal from chipless RFID tag [12] with permission from IEEE

Assuming backscattered response from the N-bit tag as shown in Fig. 4.10, the late-time part of the response can be written by

$$e(t) = \mathrm{Re}\left\{\sum_{n=1}^{N} R_{\mathrm{n}} e^{-s_{\mathrm{n}}(t-t_0)} U(t-t_0)\right\} \tag{4.53}$$

where s_{n} and R_{n} are the nth CNR and corresponding residue, respectively. The quantity t_0 is the turn-on time of the poles. In complex scatterers like airplanes, the turn-on times of the poles might not be the same due to the various scattering centers. In contrast, for simple scatterers like chipless RFID tags, the turn-on times of the CNRs are approximately the same. $U(t)$ is the step function defined as

$$U(t-t_0) = \begin{cases} 1 & t \geq t_0 \\ 0 & t < t_0 \end{cases} \tag{4.54}$$

According to Fig. 4.10, a time-window of length T_{W} is moved along the time axis incrementally by the value of T. Poles and residues of each sliding window are computed using MPM and they are indexed by T to realize a time-frequency representation. The windowed signal can be written by

$$e_{\mathrm{W}}(t) = \mathrm{Re}\left\{\sum_{n=1}^{N} R_{\mathrm{n}}^{\mathrm{T}} e^{-s_{\mathrm{n}}(t-t_0-T)} U(t-t_0)\right\} \tag{4.55}$$

In which

$$\begin{aligned} R_{\mathrm{n}}^{\mathrm{T}} &= R_{\mathrm{n}} e^{-s_{\mathrm{n}} T} \\ &= R_{\mathrm{n}} e^{-(\alpha_{\mathrm{n}} + j\omega_{\mathrm{n}}) T} \end{aligned} \tag{4.56}$$

In natural logarithmic scale, (4.56) is expressed by

$$Ln\left(\left|R_{\mathrm{n}}^{\mathrm{T}}\right|\right) = Ln(|R_{\mathrm{n}}|) - \alpha_{\mathrm{n}} T \tag{4.57}$$

Equation (4.57) indicates that in normal logarithmic scale, residues linearly decrease versus T with slope α_{n}. The real part of the poles, α_{n}, calculated from MPM are very sensitive to noise. An alternative way of calculating the damping factors of the CNRs is to find the slopes of the residues versus time in the normal logarithmic scale as (4.57) shows [12]. The simulation results presented in this section will show that the CNRs calculated from STMPM are more accurate than those from MPM. By calculating the poles and residues, three different diagrams (time-frequency, time-damping factor, and time-residue) can be used in the detection and localization of the tags.

Two important parameters in STMPM are the length of the window, T_{W}, and filtering parameter, p, defined in (4.48). The time and frequency resolution of this technique can be improved by choosing suitable values of T_{W} and p. Similar to STFT, by increasing the length of the window, the frequency resolution improves, while the time resolution deteriorates and vice versa. As seen in (4.48), by increasing the value of p, the poles with lower energy can be recognized in the presence of the high-energy poles. This ability is the significant advantage of the proposed technique over STFT and wavelet decomposition for detection of the chipless RFID tags.

Because of the direction and polarization dependency of the scattered signal, some of the poles may be excited more strongly than other poles for a specific polarization and direction of arrival. In such circumstances, by increasing the value of the filtering parameter p, the low energy poles can be extracted from the signal. Additionally, with any increase in p, the frequency resolution of the tag can be improved without increasing Tw.

To show this effect, we consider a challenging situation involving two poles with severely different energy. Ignoring the noise, we first consider a signal containing two singularity poles as

$$s(t) = R_1 e^{-\alpha_1 t} \cos(\omega_1 t + \varphi_1) + R_2 e^{-\alpha_2 t} \cos(\omega_2 t + \varphi_2) \tag{4.58}$$

Assuming $f_1 = \omega_1/2\pi = 5$ GHz, $f_2 = 6$ GHz, $\varphi_1 = \varphi_2 = \pi/4$, $\alpha_1 = \alpha_2 = $ 1e9 and $R_1 = 20R_2$, the signal and its time-frequency diagram are shown in Fig. 4.11 for different values of T_{W} and p. In Fig. 4.11b, by choosing $T_{\mathrm{W}} = 1.1$ ns and $p = 4$, both resonances can be easily detected in the time-frequency diagram. By decreasing the value of filtering parameter to $p = 2$ and keeping the length of the window $T_{\mathrm{W}} = 1.1$ ns, the second pole is not detected. For the same value of p, by increasing the length of the window, the second resonance of the signal cannot be detected. It

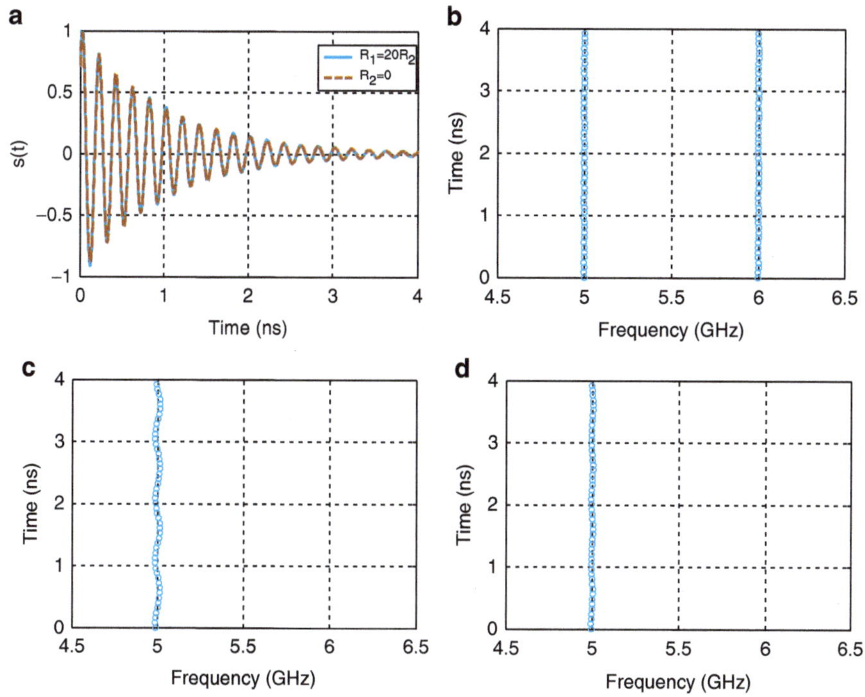

Fig. 4.11 (**a**) Time-domain signal and time-frequency response for (**b**) $T_W = 1.1$ ns, p = 4 (**c**) $T_W = 1.1$ ns, p = 2 (**d**) $T_W = 2.5$ ns, p = 2 [13] with permission from IEEE

is clearly seen that by then increasing the value of p, the weaker pole will be detected in the time-frequency diagram. In Fig. 4.12, the time-residue diagram [12] of the signal is represented in natural logarithmic scale for S_1: $R_1 = R_2 = 1$, $\alpha_1 = 2\alpha_2 = 1e9$ and S_2: $R_1 = 10R_2 = 1$, $\alpha_1 = 2\alpha_2 = 1e9$. The slopes of the lines and values of the residues at the turn-on times are proportional to the damping factor and absolute value of the residues, respectively. In S_1, the residues of poles are the same, therefore the lines start from the same point in the time-residue diagram. Since the damping factor of the second pole is half of the first pole's one, the slope of the line increases by factor of 2. By changing the residue of the second pole in the second signal, the start point of the line is changed in the time-residue diagram.

Another important parameter is the time-window length, shown by T_W. This parameter affects the time and frequency resolutions. In the cases where the resonant frequencies of the scatterer are close to each other, we need to increase T_W to distinguish the resonances. Assuming $R_1 = R_2$, $\varphi_1 = \varphi_2 = \pi/4$ and $\alpha_1 = \alpha_2 = 1e9$ and $f_1 = 5$ GHz, $f_2 = 5.5$ GHz, the time-frequency diagram of the signal is represented in Fig. 4.13 for two values of T_W. For $T_W = 1$ ns, the CNRs cannot be distinguished. At some instances in time, the resultant resonant frequency shows some misleading variations, taking on a value halfway between the two resonant frequencies.

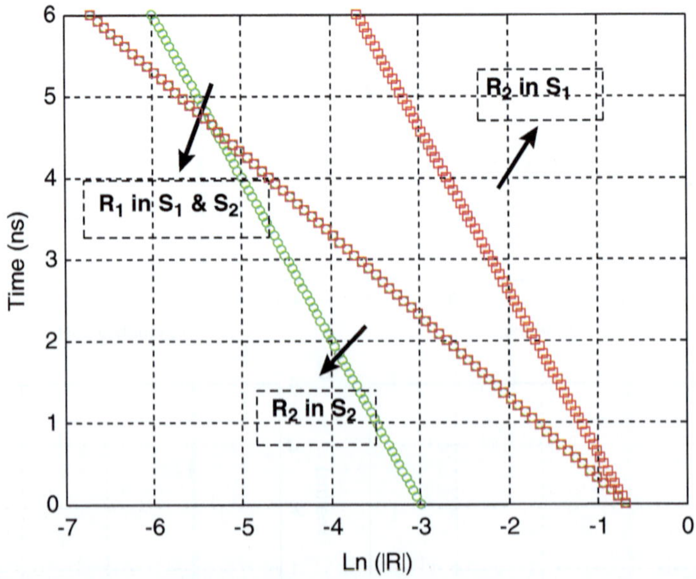

Fig. 4.12 Time-residue diagram for S_1: $R_1=R_2=1$, $\alpha_1=2\alpha_2=1e9$ and S_2: $R_1=10R_2=1$, $\alpha_1=2\alpha_2=1e9$ [13] with permission from IEEE

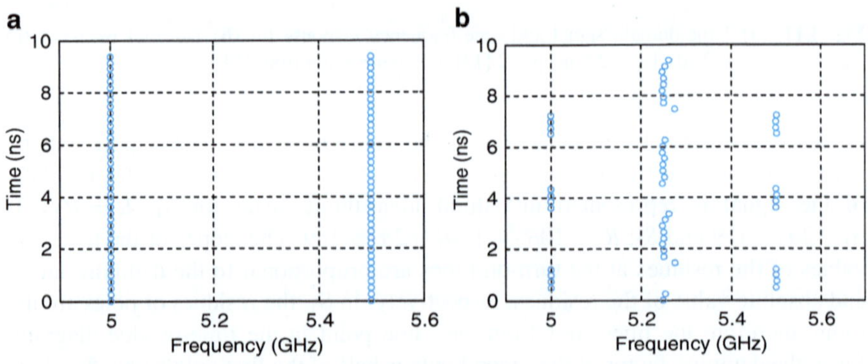

Fig. 4.13 Time-frequency diagram for (**a**) $T_W=1.8$ ns, p = 2. (**b**) $T_W=1$ ns, p = 2 [13] with permission from IEEE

Hence, any increase in T_W leads to a better resolution in the frequency domain. In addition to T_W, the parameter p affects the resolution in time and frequency domains. In order to study the effects of both T_W and p on resolution, the signal seen in (4.58) is considered with $R_1=R_2$, $\varphi_1=\varphi_2=\pi/4$ and $\alpha_1=\alpha_2=1e9$ and $f_1=5$ GHz. Assuming $\Delta=f_2-f_1$, the minimum value of T_W by which two resonances can both be successfully detected is shown in Fig. 4.14 versus Δ for different values of p. As can be seen, for larger values of p, a smaller window length is needed to distinguish closely spaced resonant frequencies. In other words, the resolution in both time and frequency domains is improved [13].

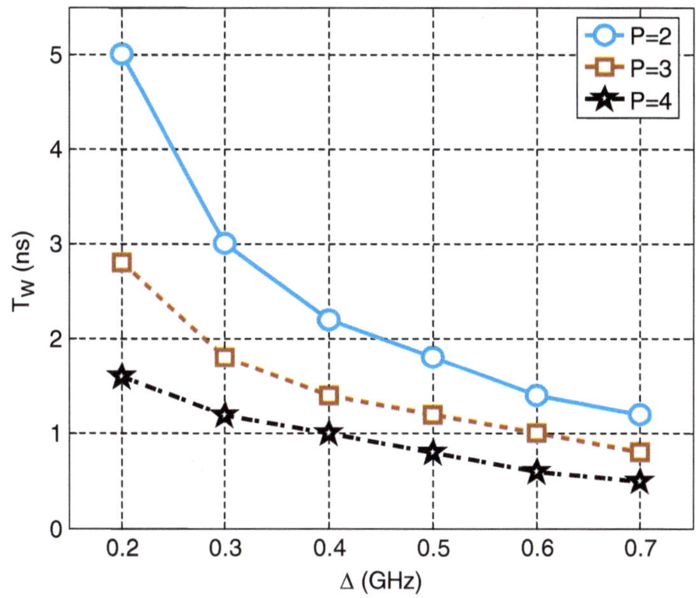

Fig. 4.14 Minimum window length required for extracting the resonances versus Δ for different values of p. $R_1 = R_2$, $\varphi_1 = \varphi_2 = \pi/4$ and $\alpha_1 = \alpha_2 = 1e9$ and $f_1 = 5$ GHz [13] with permission from IEEE

4.4.3 Application of STMPM in Scattering Process

As a time-frequency approach, various scattering mechanisms such as resonance, scattering center, and dispersion features of the scatterer can be monitored in the time-frequency diagram obtained from STMPM [13]. In some applications such as radar, the CNRs of the airplane are used as the ID for detection purposes [14–16]. In these applications, high-Q resonances are more effective for identification purposes. These high-Q resonances are mostly generated by cavity structures embedded on the scatterer. For example, the engine of the airplane makes an open-ended cavity resonator whose corresponding CNRs participate effectively in the late-time backscattered response from the airplane. Hence, the ID of the airplane can be adjusted by changing the resonant modes of the open-ended engine cavity [16]. These CNRs are usually affected by the dispersion characteristics of the structure. Figure 4.15 indicates various scattering-mechanism representations in the time-frequency diagram [13, 17]. A vertical line represents a reflection from a scattering center while a horizontal line introduces a resonance mechanism in the scattering mode.

Any slope in the time-frequency diagram (as Figs. 4.15c, d show) represents a dispersive phenomenon. In order to gather all the above mechanisms into one example, an open-ended cylindrical cavity seen in Fig. 4.16 is often considered in literature [13, 18, 19]. An incident electric field polarized in *x*-direction and propagating in $-z$ direction illuminates the cavity. The backscattered field contains

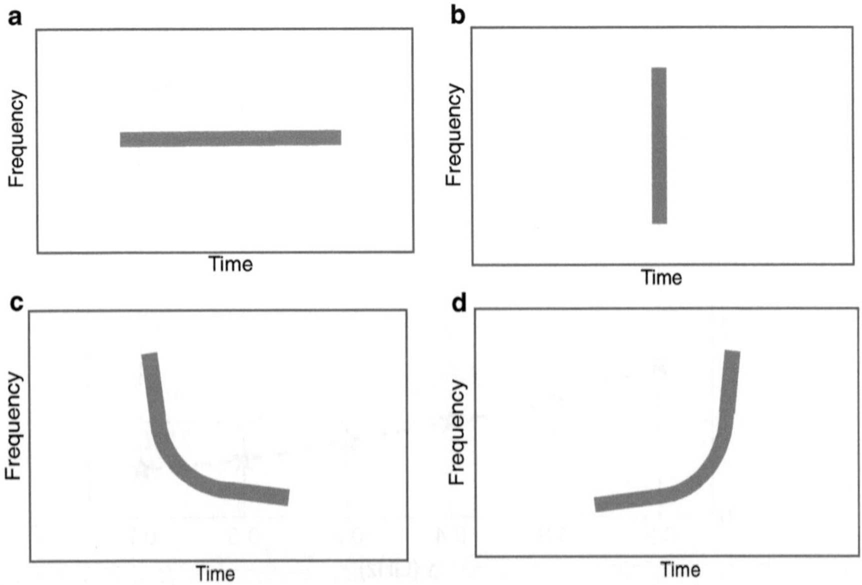

Fig. 4.15 Scattering mechanisms in time-frequency analysis. (**a**) Scattering center. (**b**) Resonant behavior. (**c**) Structural dispersion. (**d**) Material dispersion [13] with permission from IEEE

Fig. 4.16 Open-ended circular cavity excited by incident plane wave [13] with permission from IEEE

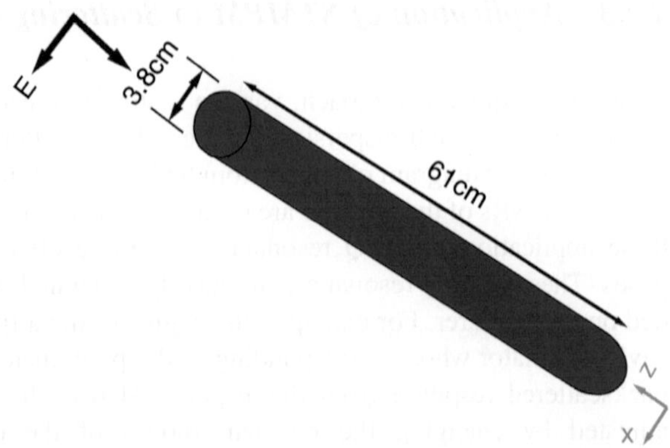

reflections from the rim and bottom of the cavity and dispersive internal resonant modes of the cavity.

As the time-domain response in Fig. 4.17 indicates, three pulse-shaped responses at $t = 1\ ns$, $t = 4.3\ ns$ and $t = 5.4\ ns$ are due to the specular reflections from the rim, and the external and internal back of the cavity, respectively. The time-frequency diagram of the signal is depicted in Fig. 4.18 using STMPM and STFT. The parameters of STMPM are chosen as $T_w = 0.4\ ns$ and $p = 2$. The incident pulse covering the frequencies from 10 MHz to 20 GHz excites the internal modes of the cavity. Each internal mode is associated with numerous resonances which are

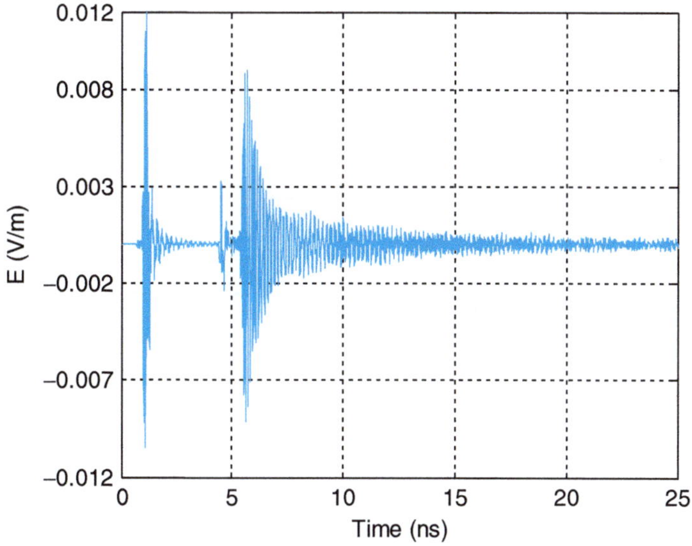

Fig. 4.17 Time-domain backscattered field from the waveguide cavity [13] with permission from IEEE

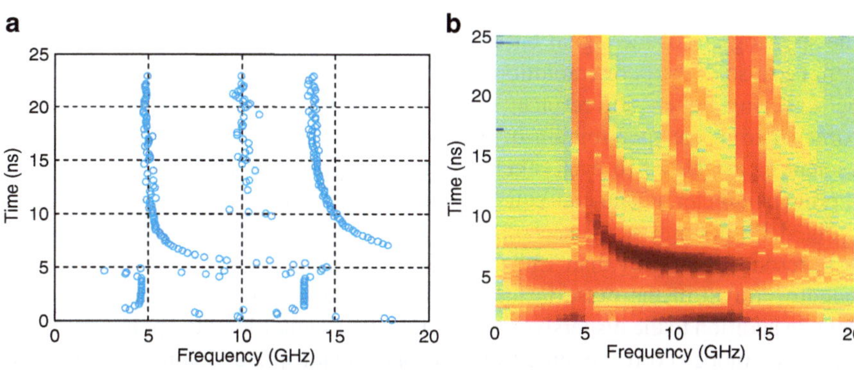

Fig. 4.18 Time-frequency diagram of the backscattered signal from the cylinder based on (**a**) STMPM and (**b**) STFT [13] with permission from IEEE

mainly affected by the mode itself with very negligible effect of the external structure. There is a cluster of high-Q resonances slightly above the modal cutoff frequency, and many more resonances with lower Q at higher frequencies. The effect of these high-order resonances are dominant at earlier times of the response. According to figure, there are two scattering centers and three resonant modes. The modes with cut-off frequencies $f = 5$ GHz and $f = 13.8$ GHz have been excited more strongly than the mode with cut-off resonance at $f = 9.9$ GHz. There are some poles parallel to the frequency axis located around $t = 10$ ns which emanate from the second roundtrip travel of the pulse inside the waveguide cavity.

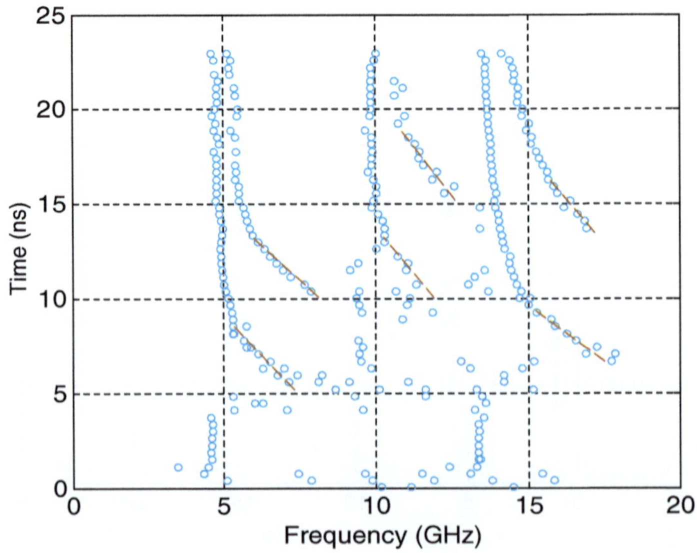

Fig. 4.19 Time-frequency diagram of the scattered field for $T_W = 1$ ns, $p = 4$ [13] with permission from IEEE

The spectrogram of the signal based on STFT is shown in Fig. 4.18b. As can be seen, because of the poor resolution of STFT in both frequency and time domains, the scattering centers and resonant frequencies of the cavity cannot be accurately extracted from the spectrogram of the signal. This is especially severe when the resonances are closer to each other. By increasing the length of the window and filtering parameter, p, and consequently increasing the frequency resolution as seen in Fig. 4.19, not only is the second mode clearly visible, but there also exist some extra resonances higher than 13.8 GHz. These frequencies result from the second roundtrip travel of the pulse inside the cavity, which shows itself at later times. Here, in addition to the dispersive modes of the first roundtrip of the pulse inside the cavity, the modes corresponding to the second roundtrip are visible with different turn-on times which is exactly matched with the electrical length of the cavity. The dispersion characteristics of the modes have been highlighted with dashed lines. Another advantage of the proposed technique is that the local resonant frequencies of the scatterer are illustrated by discrete poles in the time-frequency diagram rather than continuous colors in wavelet and STFT. Figure 4.20 shows the time-frequency diagram of the signal based on the proposed technique for SNRs of 20 and 10 dB. For lower values of SNR, we need to decrease the value of p in order to avoid the presence of poles originating from noise.

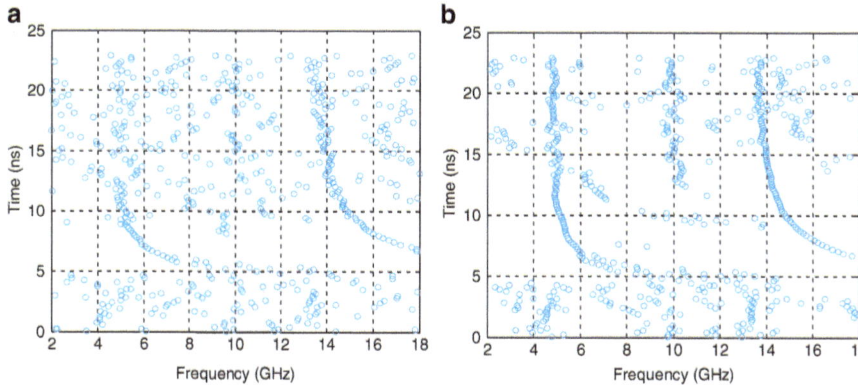

Fig. 4.20 Time-frequency diagram of the signal with (**a**) SNR = 10 dB, $T_W = 0.8$ ns, p = 2 and (**b**) SNR = 20 dB. $T_W = 0.8$ ns, p = 4 [13] with permission from IEEE

4.4.4 Application of STMPM in Chipless RFID Tags

Since the received signal at the antenna ports incorporates noise, the accuracy of the detection technique in decoding the IDs of the tag is very important. As shown in Chap. 3, the damping factors of the CNRs are very sensitive to environmental conditions. The damping factors are also very sensitive to noise. For signal-to-noise ratios less than 20 dB, the damping factors obtained from MPM are not accurate [12]. Compared to the damping factors, the resonant frequencies are more robust against the environmental issues and noise. Figure 4.10 shows the noiseless backscattered signal from a 3-bit tag. The fundamental resonant frequencies of the tag are located at $f_1 = 5.2$ GHz, $f_2 = 7.1$ GHz, and $f_3 = 8.6$ GHz. The time-frequency and time-damping factor diagrams of the tag are depicted in Fig. 4.21. The length of the sliding window is $T_W = 0.9$ ns and the filtering parameter is chosen as $p = 2$. As Fig. 4.21 shows, by sliding the window in the early-time response of the tag, all components from low to high frequencies are seen in the time-frequency diagram. Because the early-time response is a pulse-shaped signal proportional to the incident pulse, it includes all the frequency components covered in the bandwidth of the incident pulse. At turn-on times, the poles in the early-time converge to the CNRs in the late-time part. In this example, the turn-on time of the CNRs is at t = 0.65 *ns*. Due to the exponentially decaying nature of the late-time signal, the accuracy of the extracted poles decreases with time. The extracted damping factor shows some variations along the time axis. In Fig. 4.22, the time-residue of the signal is shown. The slope of the lines is equal to the damping factor of the CNRs. As it shows, the extracted data has lower variations along the lines, which ensures the accuracy of this technique compared to the matrix-pencil method.

The pole-diagram of the 3-bit tag is shown in Fig. 4.23. It shows the resonant frequencies and damping factors of the tag in one diagram. In order to study the

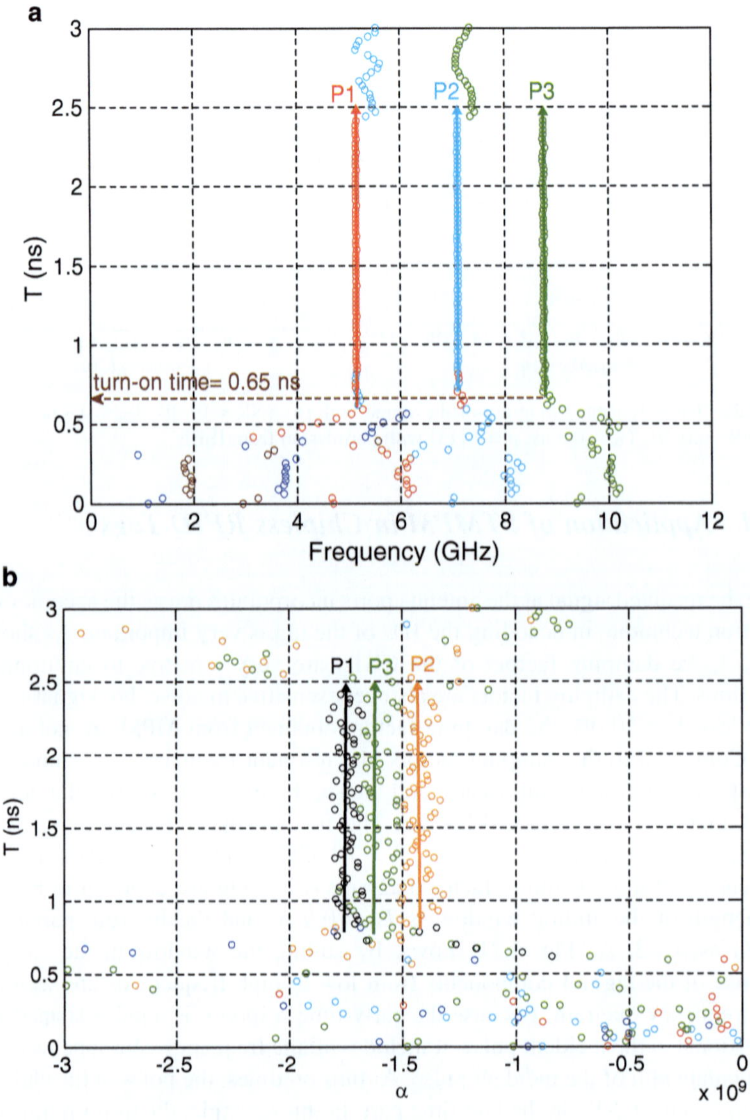

Fig. 4.21 (**a**) Time-frequency and (**b**) time-damping factor diagrams of the scattered signal from 3-bit chipless RFID tags [12] with permission from IEEE

effect of noise in extracted poles from STMPM, the time-frequency, time-damping factor, and time-residue diagrams of the backscattered signal are depicted in Figs. 4.24, 4.25, and 4.26 for SNR = 15 dB. For lower values of SNR, the severity of the variations in damping factor increases with time, which compromises accuracy. In contrast, the accuracy of the results based on the time-residue diagram

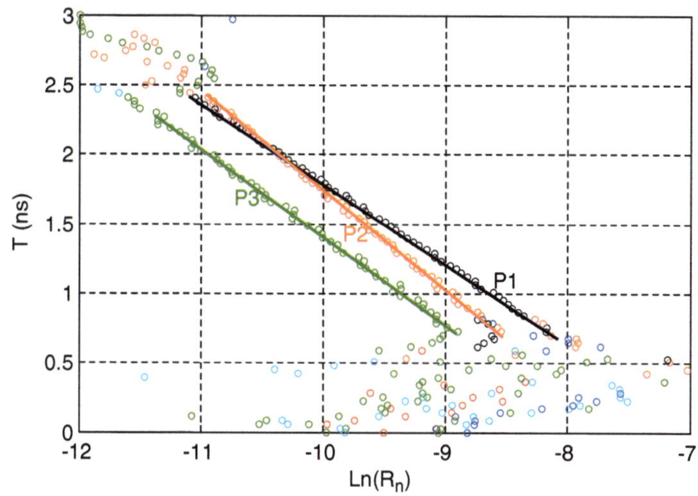

Fig. 4.22 Magnitude of the residues of dominant poles tracked as the window moves [12] with permission from IEEE

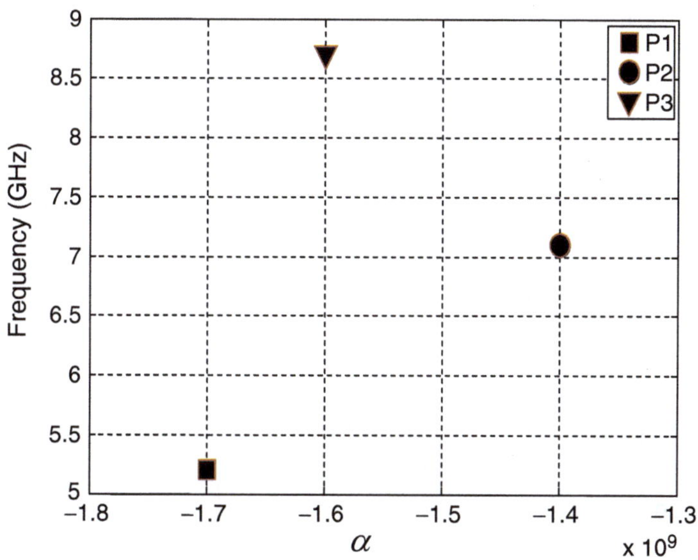

Fig. 4.23 Pole diagram of the three-bit tag [12] with permission from IEEE

seen in Fig. 4.26 is improved. For later times ($t > 1.6$ ns), since the power of the signal decreases while the power of noise does not change, by decreasing the SNR, the accuracy of the results diminishes. As an example, the unstable poles in Fig. 4.26 at times later than 1.6 ns are due to noise. The simulated results confirm

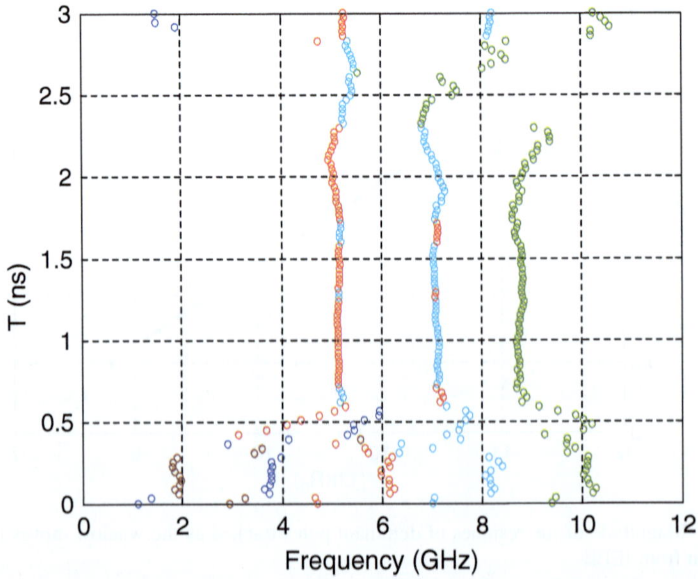

Fig. 4.24 Variation of imaginary part of the poles with sliding time for CNR = 15 dB [12] with permission from IEEE

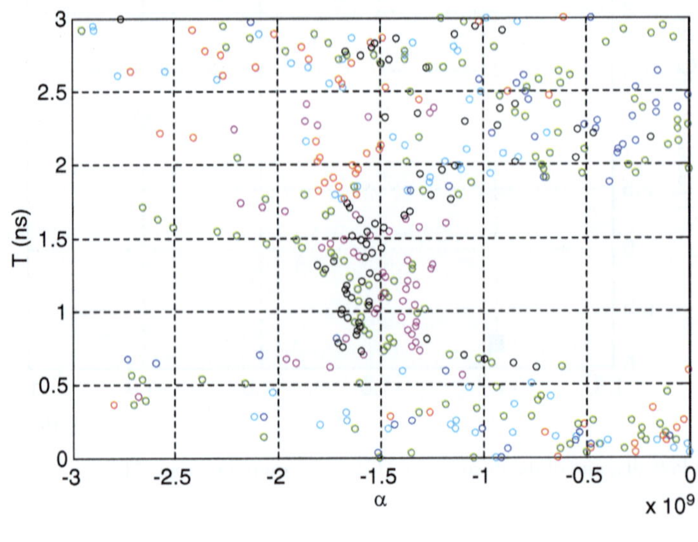

Fig. 4.25 Variation of real part of the poles with sliding time for CNR = 15 dB [12] with permission from IEEE

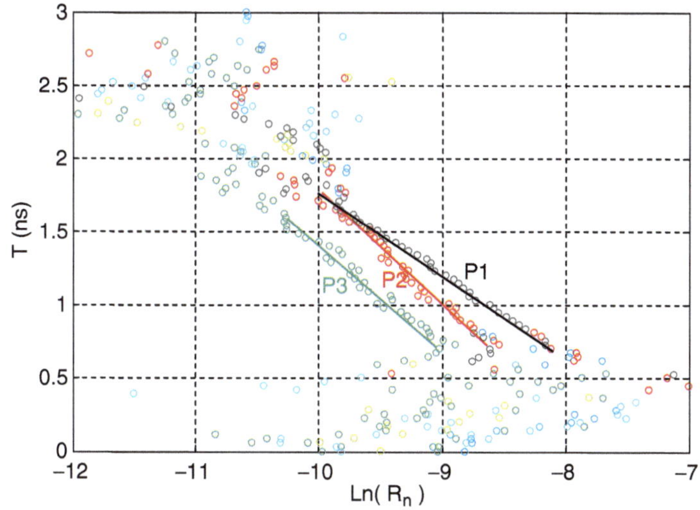

Fig. 4.26 Estimated variation of residues versus sliding time [12] with permission from IEEE

Table 4.1 Percentage error of estimating of real and imaginary parts of the dominant poles of the tag calculated from direct matrix pencil method (MPM) and short-time-matrix-pencil-method (STMPM) [12] with permission from IEEE

				α_1		α_2		α_3	
SNR	ω_1	ω_2	ω_3	MPM	STMPM	MPM	STMPM	MPM	STMPM
15	0.39	0.31	0.36	9.14	7.18	12.7	7.1	11.1	6.78
10	0.71	0.68	0.82	17	15.3	22	15.1	21.6	14.6
5	2.85	2.98	3.2	28.4	24.6	31.5	26.8	48	29

the improved accuracy of the CNRs calculated from the STMPM technique compared to MPM. Table 4.1 presents the average error in estimating the real and imaginary parts of the poles for 50 different sets of noisy data with a specific SNR value. As the table shows, the proposed method gives significantly more accurate results for damping factors than MPM.

References

1. Stutzman WL, Thiele GA (1998) Antenna theory and design. Wiley, New York, NY
2. Reed JH (2005) An introduction to ultra wideband communication systems. Prentice Hall PTR, Upper Saddle River, NJ
3. Licul S, Davis WA (2005) Unified frequency and time-domain antenna modeling and characterization. IEEE Trans Antennas Propag 53:2882–2888

4. Qian S, Dapang C (1999) Joint time-frequency analysis. IEEE Signal Process Mag 16:52–67
5. Poularikas AD (2000) Transforms and applications handbook. CRC Press, Boca Raton, FL
6. Chen VC, Ling H (2002) Time-frequency transforms for radar imaging and signal analysis. Artech House, Boston, MA
7. VanBlaricum M, Mittra R (1978) Problems and solutions associated with Prony's method for processing transient data. IEEE Trans Antennas Propag 26:174–182
8. Sarkar TK, Pereira O (1995) Using the matrix pencil method to estimate the parameters of a sum of complex exponentials. IEEE Antennas Propag Mag 37:48–55
9. Rothwell E, Nyquist DP, Chen YF, Drachman B (1985) Radar target discrimination using the extinction-pulse technique. IEEE Trans Antennas Propag 33:929–937
10. Woojin L, Sarkar TK, Hongsik M, Salazar-Palma M (2012) Computation of the natural poles of an object in the frequency domain using the Cauchy method. IEEE Antennas Wirel Propag Lett 11:1137–1140
11. Licul S (2004) Ultra-wideband antenna characterization and measurements. Ph.D. dissertation, Virginia Tech., Blacksburg, VA
12. Rezaiesarlak R, Manteghi M (2013) Short-time matrix pencil method for chipless RFID detection applications. IEEE Trans Antennas Propag 61:2801–2806
13. Rezaiesarlak R, Manteghi M (2015) On the Application of Short-Time Matrix Pencil Method for Wideband Scattering from resonant structures. IEEE Trans Antennas Propag 63(1)
14. Moffatt DL, Mains RK (1975) Detection and discrimination of radar targets. IEEE Trans Antennas Propag 23:358–367
15. Li Q, Ilavarasan P, Ross JE, Rothwell EJ, Chen YF, Nyquist DP (1998) Radar target identification using a combined early-time/late-time E-pulse technique. IEEE Trans Antennas Propag 46:1272–1278
16. Chauveau J, De Beaucoudrey N, Saillard J (2010) Resonance behavior of radar targets with aperture: example of an open rectangular cavity. IEEE Trans Antennas Propag 58:2060–2068
17. Chen V, Hao L (1999) Joint time-frequency analysis for radar signal and image processing. IEEE Signal Process Mag 16(2):81–93
18. Moghaddar A, Walton KE (1993) Time-frequency analysis of scattering from waveguide cavities. IEEE Trans Antennas Propag 41(5):677–680
19. Ram SS, Hao L (2007) Application of the reassigned joint time-frequency transform to wideband scattering from waveguide cavities. IEEE Antenna Wireless Propag Lett 6:580–583

Chapter 5
Detection, Identification, and Localization in Chipless RFID Systems

5.1 Introduction

The scattered signal from chipless RFID tags is received by the receiving antenna and transferred to the reader. In the reader, some information such as the location and ID of the tags should be extracted from the received signal. In applications where multiple tags are present in the reader area, we need to first distinguish the tag's responses from the responses of the background objects around the tag. After the detection process, the locations and IDs of the tags should be extracted, the process of which comprises the localization and identification process in the reader.

When multiple tags are present in the reader area, we need an anti-collision algorithm to separate the IDs of the tags. In [1], fractional Fourier transform as a joint time-frequency transformation is employed to separate the responses from multiple tags. Assuming the interrogating signal is a chirp signal, the received signal is a summation of the reflected signals from the tags. Since the tags are located at different distances from the antennas, the received chirps have different delays. Various time-frequency approaches, mentioned in Chap. 4, such as STFT, wavelet or adaptive techniques can be used to separate the chirped signals in the time-frequency diagram. The drawback of these techniques is their poor resolution or cross-term interference when the tags are located in close proximity to each other. In [1], by applying fractional Fourier transform, the time-frequency plane is rotated by an optimum angle to minimize the overlap region between chirp signals. The number of the tags can be estimated from the number of spikes observed in the transformed signal. Although the positions of the tags can be obtained using this approach, their IDs cannot be easily separated by applying fast Fourier transform (FFT) to the windowed signal, as mentioned in [1]. Because by adjusting the window exactly after the turn-on time of a specific tag, the frequency components of the considered tag and all pre-illuminated tags are included in the windowed signal, which complicates the identification process . This is more severe when the tags are close to each other.

R. Rezaiesarlak, M. Manteghi, *Chipless RFID*, DOI 10.1007/978-3-319-10169-9_5

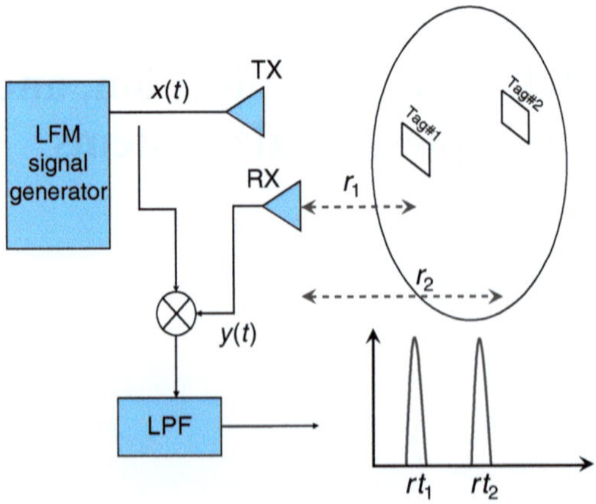

Fig. 5.1 Block diagram of Anti-collision algorithm [2] with permission from IEEE

In [2], the idea of frequency modulated continuous wave (FMCW) radar has been used as a collision avoidance algorithm. In the proposed technique, they have utilized a linear chirp signal as an illuminating signal. The block diagram of the system is depicted in Fig. 5.1 [2]. A transmitting antenna transmits the following signal:

$$x(t) = A\sin\left(2\pi f_0 t\right) \tag{5.1}$$

Assuming two tags are present in the reader area, the interrogating signal illuminates the tags located at different distances from the receiving antennas. The received signal contains the reflections from the tags at the frequency f_0 with different amplitudes and phases as

$$y_1(t) = A_1\sin\left(2\pi f_0 t + \varphi_1\right) \tag{5.2a}$$
$$y_2(t) = A_2\sin\left(2\pi f_0 t + \varphi_2\right) \tag{5.2b}$$

At the times that the antenna receives the delayed signals reflected from the tags, the transmitting signals are

$$x_1(t) = A\sin\left(2\pi f_1 t\right) \tag{5.3a}$$
$$x_2(t) = A\sin\left(2\pi f_2 t\right) \tag{5.3b}$$

where $f_1 = f_0 + rt_1$ and $f_2 = f_0 + rt_2$ are the frequencies of the transmitting signals at $t = t_1$ and $t = t_2$. Quantities $\Delta t_1 = t_1 - t_0$ and $\Delta t_2 = t_2 - t_0$ are the delays of the received signals from the tags and r is the chirp rate. By mixing the received signals with the transmitting ones at $t = t_1$ and $t = t_2$ and suppressing the high-frequency components, the resultant signals after low-pass filtering are

$$z_1(t) = \frac{AA_1}{2}\cos\left(2\pi r t_1 t - \varphi_1\right) \tag{5.4a}$$

$$z_2(t) = \frac{AA_2}{2}\sin\left(2\pi r t_2 t - \varphi_2\right) \tag{5.4b}$$

As (5.4a and 5.4b) shows, the received signals from the tags are separated in the frequency domain. Similar to the aforementioned technique based on fractional Fourier transform, the IDs of the tags cannot be easily separated from the windowed signals.

As mentioned in Chap. 3, the ID of the tag is generated by arranging the resonant frequencies of the structure. Because the ID is included in the late-time response of the received signal, a time-frequency representation is needed in order to extract the ID of the tags from the late-time part. In cases where multiple tags are in the main beam of the antenna, another dimension for space should be added to the representation diagram. In this chapter, a space-time-frequency representation is introduced by which all the required information of the tags can be obtained. In order to improve the performance of the system, one must consider the resolution of the applied algorithm in space, time, and frequency domains. After introduction of the new space-time-frequency algorithm for the detection, identification, and localization of a chipless RFID tag, the effects of various parameters on the resolution in space, time, and frequency domains are studied.

5.2 Space-Time-Frequency Anti-collision Algorithm for Identifying Chipless RFID Tags

In Fig. 5.2, the schematic view of multiple tags presented in the main beam of the reader antenna is depicted. As it shows, the received backscattered signal is composed of the reflected signals from each tag. Assuming the configuration in Fig. 5.2, the scattered field can be written in the Laplace-domain as (5.5).

$$\mathbf{E}^s(\mathbf{r}) = \int \overleftrightarrow{\mathbf{G}}_{\mathrm{e}}\left(\mathbf{r}, \mathbf{r}'; s\right) \cdot \mathbf{J}_{\mathrm{s}}\left(\mathbf{r}'\right) ds' \tag{5.5}$$

The source points are presented in the primed coordinate and the observation points are presented using unprimed coordinates. Quantity $\overleftrightarrow{\mathbf{G}}_e$ is the electric dyadic Green's function [4, 5] and $\mathbf{J}_\mathrm{s}$ is the surface current density induced on the scatterer. Here, the reader area is assumed as a scattering medium with tags as the scattering centers. For the case of multiple tags, the current density can be written as the summation of the currents on the tags as

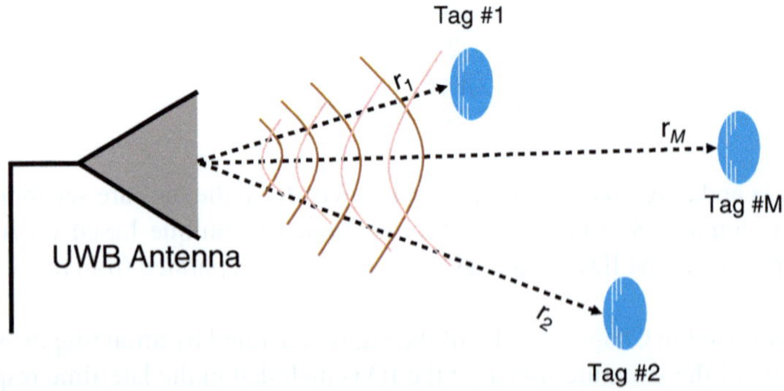

Fig. 5.2 Multiple chipless RFID tags present in the reader zone [3] with permission from IEEE

$$\mathbf{J}_s\left(\mathbf{r}'\right)=\sum_{m=1}^{M}\mathbf{J}_{sm}\left(\mathbf{r}'-\mathbf{r}'_m\right) \tag{5.6}$$

in which M is the number of tags, and $\mathbf{r}'_m$ and $\mathbf{J}_{sm}$ are the location and induced current on the mth tag, respectively. Backscattered fields from the tags can be deduced from the electric field integral equation (EFIE) as

$$\int \overleftrightarrow{\mathbf{G}}_e\left(\mathbf{r},\mathbf{r}';s\right)\cdot\mathbf{J}_s\left(\mathbf{r}-\mathbf{r}'\right)ds' = -\hat{t}\cdot\mathbf{E}^{\mathbf{inc}}(\mathbf{r}) \ \ \forall\mathbf{r}\in S \tag{5.7}$$

in which $\mathbf{E}^{\text{inc}}$ is the incident electric field which emanates from the antenna, $\hat{t}$ represents the unit vector tangential to the tag surface, and $s=\alpha+j\omega$ is the complex frequency. On the other hand, based on singularity expansion method (SEM) [6], the current in (5.6) and (5.7) can be expanded by a series of complex natural resonances as [7]

$$\mathbf{J}_s\left(\mathbf{r}';s\right)=\sum_{m=1}^{M}\sum_{n=-N_m}^{N_m}\frac{\overleftrightarrow{\mathbf{A}}_m^{(n)}\cdot\mathbf{J}_{sm}^{(n)}\left(\mathbf{r}'-\mathbf{r}'_m\right)}{\left(s-s_m^{(n)}\right)}+\sum_{m=1}^{M}\mathbf{J}_m\left(\mathbf{r}'-\mathbf{r}'_m;s\right) \tag{5.8}$$

where $\mathbf{J}_{sm}^{(n)}$ and $s_m^{(n)}$ are the nth natural-mode current and pole of the mth tag. Compared to complex natural resonances, the residue coefficients, $\overleftrightarrow{\mathbf{A}}_m^{(n)}$, are aspect-dependent, depending strongly on the polarization and incident angle. The last summation in (5.8) contains the entire-domain function including the early-time response from the scatterers [7–9]. The early-time response originates from the scattering centers of the scatterer [10–14]. By inserting (5.8) in (5.7) and applying the method of moments, the induced currents on the tags can be obtained. Assuming the frequency band of operation covers all the natural resonances of the tags, the backscattered signal from induced current on the tags can be written in time domain as

$$\mathbf{e}^{\mathrm{s}}(\mathbf{r},t)=\sum_{m=1}^{M}\left[\mathbf{e}_{\mathrm{m}}(\mathbf{r},t)+\mathrm{Re}\left(\sum_{n=1}^{N_{\mathrm{m}}}\mathbf{R}_{\mathrm{n}}^{(\mathrm{m})}e^{-s_{\mathrm{n}}^{(\mathrm{m})}(t-t_{\mathrm{m}})}\right)\right] \tag{5.9}$$

in which t_{m} represents the turn-on time of the mth tag. It is assumed that CNRs of each tag have the same turn-on time. For complex scatterers, this assumption might not be accurate. According to Altes' model [10] mentioned in Chap. 2, the early-time response can be approximated by a series of pulse responses as

$$\mathbf{e}_{\mathrm{m}}(\mathbf{t})=\mathbf{e}^{\mathrm{inc}}(\mathbf{r},t)*\sum_{p=-\infty}^{+\infty}a_{\mathrm{mp}}\delta^{(\mathrm{p})}(t-t_{\mathrm{m}}) \tag{5.10}$$

where t_m is the delay time equal to the roundtrip time between the antenna and the mth tag. The impulse response of the mth tag in (5.10) is summed over the integrals and derivatives of the Dirac-delta function. Here, the negative and positive value of p refers to the pth integral and derivative of the delta function. Therefore, (5.9) can be written in Laplace-domain as

$$\begin{aligned}\mathbf{E}^{\mathrm{s}}(\mathbf{r},s)&=\sum_{m=1}^{M}\left[\sum_{p=-\infty}^{+\infty}a_{\mathrm{mp}}s^{p}\mathbf{E}^{\mathrm{inc}}(\mathbf{r},s)e^{-st_{\mathrm{m}}}+\sum_{n=-N_m}^{N_m}\frac{\mathbf{R}_{\mathrm{m}}^{(\mathrm{n})}}{s-s_{\mathrm{m}}^{(\mathrm{n})}}\right]\\&=\sum_{m=1}^{M}\left[\mathbf{A}_{\mathrm{m}}(\mathbf{r},s)e^{-st_{\mathrm{m}}}+\sum_{n=-N_m}^{N_m}\frac{\mathbf{R}_{\mathrm{m}}^{(\mathrm{n})}}{s-s_{\mathrm{m}}^{(\mathrm{n})}}\right]\end{aligned} \tag{5.11}$$

in which the early-time summation is summarized by $\mathrm{A_m}$. Comparing (5.9) with (5.11), there is a duality between late-time response in time domain and the early-time response in Laplace domain [12]. By applying STMPM to (5.9), the complex resonances of the tags and their residues are found at each snapshot of time. By shifting the sliding window by T in the late-time region, the backscattered signal can be written as

$$\mathbf{e}_{\text{late-time}}^{\mathrm{s}}(\mathbf{r},t)=\mathrm{Re}\left(\sum_{m=1}^{M}\sum_{n=1}^{N_{\mathrm{m}}}\mathbf{R}_{\mathrm{m}}^{\mathrm{T,n}}e^{-s_{\mathrm{m}}^{(\mathrm{n})}(t-t_{\mathrm{m}}-T)}\right) \tag{5.12}$$

in which

$$\mathbf{R}_{\mathrm{m}}^{\mathrm{T,n}}=\mathbf{R}_{\mathrm{m}}^{(\mathrm{n})}e^{-\left(\alpha_{\mathrm{m}}^{(\mathrm{n})}+j\omega_{\mathrm{m}}^{(\mathrm{n})}\right)T} \tag{5.13}$$

In normal logarithmic scale, (5.13) can be expressed as

$$Ln\left(\left|\mathbf{R}_{\mathrm{m}}^{\mathrm{T,n}}\right|\right)=Ln\left(\left|\mathbf{R}_{\mathrm{m}}^{\mathrm{n}}\right|\right)-\alpha_{\mathrm{m}}^{(\mathrm{n})}T \tag{5.14}$$

As can be seen in (5.14), the residues linearly decrease versus T with slope $\alpha_{\mathrm{m}}^{(\mathrm{n})}$. By applying Narrow-frequency matrix pencil method (dual of STMPM) to the

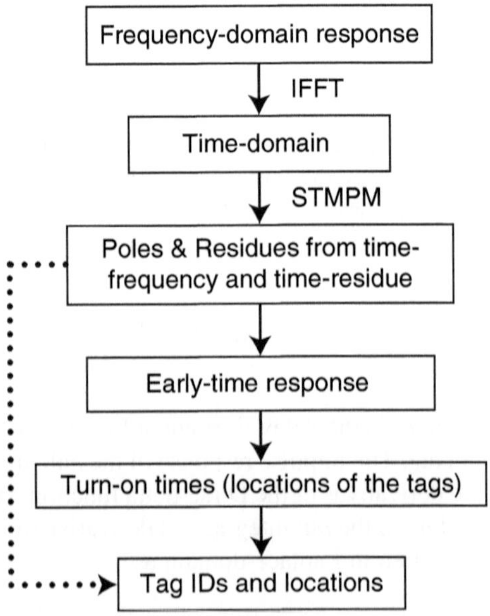

Fig. 5.3 Flowchart of the proposed anti-collision algorithm [3] with permission from IEEE

frequency response in (5.11), a space-frequency representation is obtained in which the scattering centers (here the tags) of the response are depicted versus frequency. In this diagram, the resonant frequencies of the tags in the late time are converted to some unstable poles. In practical applications limited to the frequency band of 3.1–10.6 GHz, when some dense multi-bit tags exist in the reader zone, the resonances of the tag in the frequency domain perturb the early-time response. In these cases, by inserting the poles and related residues in (5.11), the early-time part of the signal is found as

$$\mathbf{E}^{s}_{\text{early-time}}(\mathbf{r}, s) = \sum_{m=1}^{M} \mathbf{A}_{\text{m}}(\mathbf{r}, s)e^{-st_{\text{m}}} \tag{5.15}$$

By applying NFMPM to (5.15), the delay times (t_{m}) are accurately obtained. The flowchart of the proposed algorithm is illustrated in Fig. 5.3. The aforementioned technique is generalized and can be used for any multi-resonance tag schemes.

For better illustration, Fig. 5.4a shows a scenario in which two single-bit tags are illuminated by a plane wave. In the first case, the resonant frequencies of the tags are assumed at $f_1 = 7.8$ GHz and $f_2 = 9.8$ GHz and the tags are located $R = 20$ cm away from each other. Each tag is characterized by a complex natural resonance (CNR) which is created by inserting a quarter-wavelength slot on the tag surface [15]. The approach is applicable for other types of tag including slot, transmission line, or spiral resonators [16–21]. The time-domain response is shown in Fig. 5.5a. The early-time and late-time responses of the tags are illustrated by two different lines. The sliding time and window length are shown by T and $T_{\text{W}} = 0.5\ ns$. The

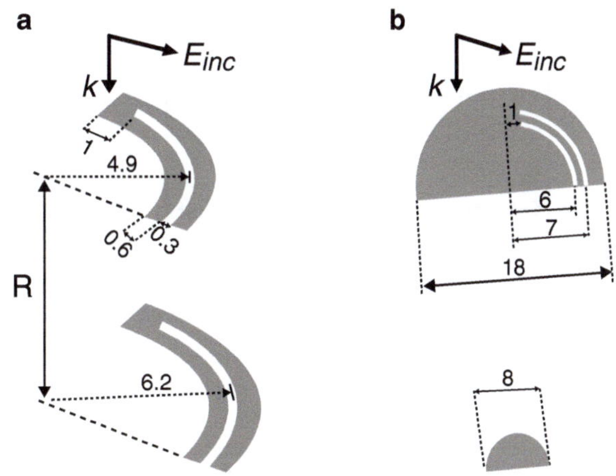

Fig. 5.4 (**a**) Two single-bit tags and (**b**) two 2-bit tags spaced by R are illuminated by a plane wave. Units in mm [3] with permission from IEEE

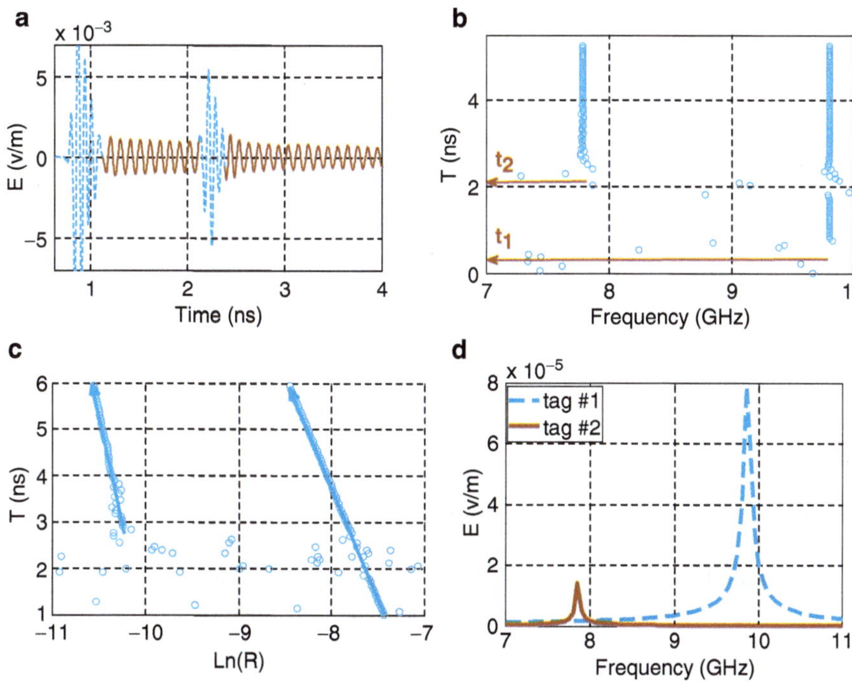

Fig. 5.5 (**a**) Time-domain backscattered signal from two tags spaced by $R = 20$ cm. (**b**) Time-frequency representation of the signal by applying STMPM with $T = 0.5$ ns. (**c**) Time-residue diagram of the signal. (**d**) Separated responses of the tags in frequency-domain [3] with permission from IEEE

time-frequency representation of the response is shown in Fig. 5.5b. According to the figure, the unstable poles in the early-time part converge to the CNRs at turn-on times (t_1 and t_2). Then, after $t_2 = 2.4$ ns, the response contains the backscattered signal from both tags; whereas before t_2, the backscattered signal contains just the

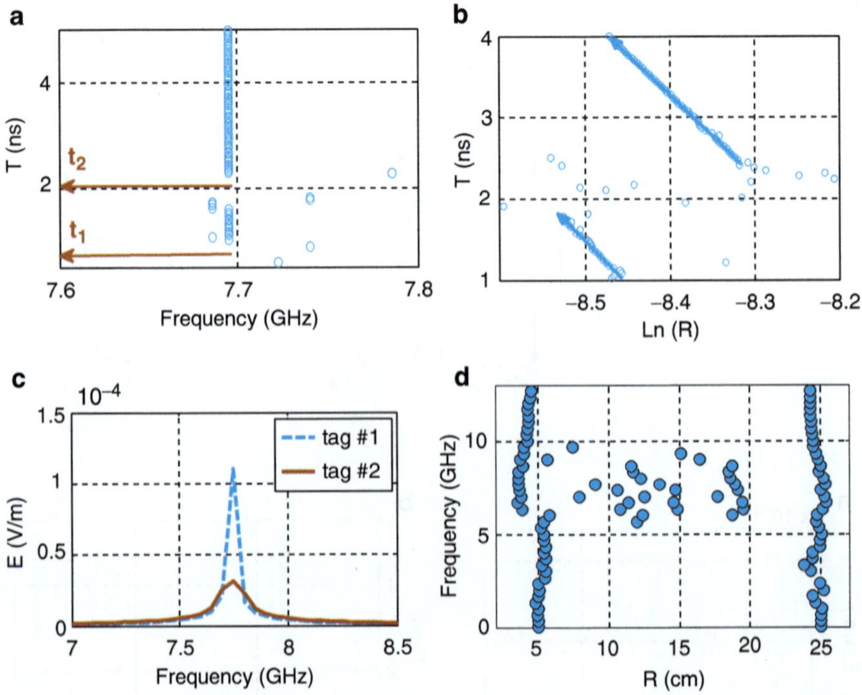

Fig. 5.6 (**a**) Time-frequency and (**b**) time-residue representation of the signal by applying STMPM with T = 0.5 ns. (**c**) Separated responses of the tags in frequency-domain. (d) Space-frequency response after SFMPM [3] with permission from IEEE

response from the first tag. In contrast to the time-frequency representation of the signal, in the time-residue diagram of Fig. 5.5c, two poles can be easily distinguished. The pole with turn-on time $t_1 = 1$ ns is associated with tag #1 and likewise, the pole with turn-on time $t_2 = 2.4$ ns is related to the second tag. According to (5.14), the slopes of the lines in Fig. 5.5c are equal to the damping factor of the poles. By reconstructing the backscattered signal from the tags in the time domain, the contribution of each tag in the late-time response of the received signal is shown in Fig. 5.5d. As can be seen, the amplitude of the backscattered signal from the first tag is higher than that of the second tag. The amplitudes at the resonant frequencies are directly proportional to the residues of the poles. In addition, the frequency response of each tag is depicted separately which simplifies the identification process.

As another example, two similar single-bit tags are considered in Fig. 5.4a resonating at $f_1 = 7.7$ GHz. The time-frequency and time-residue representations of the signal are shown in Fig. 5.6a, b, respectively. Two turn-on times of the resonators are shown in Fig. 5.6a. By plotting the residues versus sliding time, there is a jump in the residue at the second turn-on time. This jump comes from the resonant frequency of the second tag. Assuming the pole of the tags as $s_1 = \alpha_1 + j\omega_1$

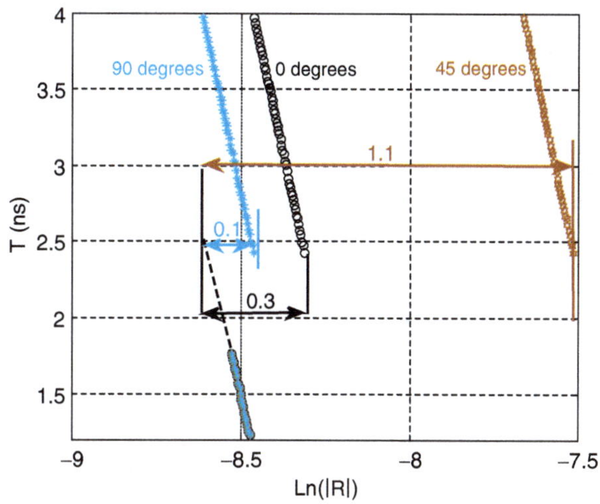

Fig. 5.7 Time-residue diagram for different polarizations [3] with permission from IEEE

and the residues of the first and second tag's poles as R_1 and R_2, then the backscattered signal after t_2 can be written as

$$\begin{aligned} & R_1 e^{-(\alpha_1+j\omega_1)(t-t_1)} + R_2 e^{-(\alpha_1+j\omega_1)(t-t_2)} \\ & \quad = e^{-(\alpha_1+j\omega_1)t}\left(R_1 e^{(\alpha_1+j\omega_1)t_1} + R_1 e^{(\alpha_1+j\omega_1)t_2}\right) \end{aligned} \tag{5.16}$$

Hence, the residue of the single pole R_1 before t_2 becomes the term in the parentheses in right-hand side of (5.16) after t_2. Since the poles of the tags have the same damping factors, the slopes of the lines are the same. The late-time frequency response of the tags is shown in Fig. 5.6c. Although they share the same resonant frequency and quality factor, they have different amplitudes at the resonant frequency. By applying NFMPM to the total frequency-domain response of the tags (without subtracting the late-time response), the locations of the tags are depicted in Fig. 5.6d in a space-frequency diagram. There are some unstable poles (between $R = 5$ cm to $R = 25$ cm) associated with the resonances of the tags. These poles can be suppressed by subtracting the late-time response from the total response in (5.11). In Fig. 5.7, the time-residue of the backscattered signal from the tags is depicted for different polarizations of tag. As can be seen when the second tag is rotated by 45° along its axis, the electric field of the incident wave is perpendicular to the slot's length and it excites the tag's pole more effectively. According to Fig. 5.7, the residue of the first tag does not change while the residue of the second tag shifts proportional to the polarization of the second tag. When the slot's length is in parallel with the incident electric field (90° rotation), the excited residue of the second tag is small and as can be seen in Fig. 5.7, it produces a small shift at the second turn-on time.

In the third example as shown in Fig. 5.4b, two 2-bit tags are illuminated by an incident plane wave. The first illuminated tag represents two resonant frequencies at $f_1 = 5.3$ GHz and $f_2 = 7.1$ GHz (ID = 11) and the second tag has no resonant

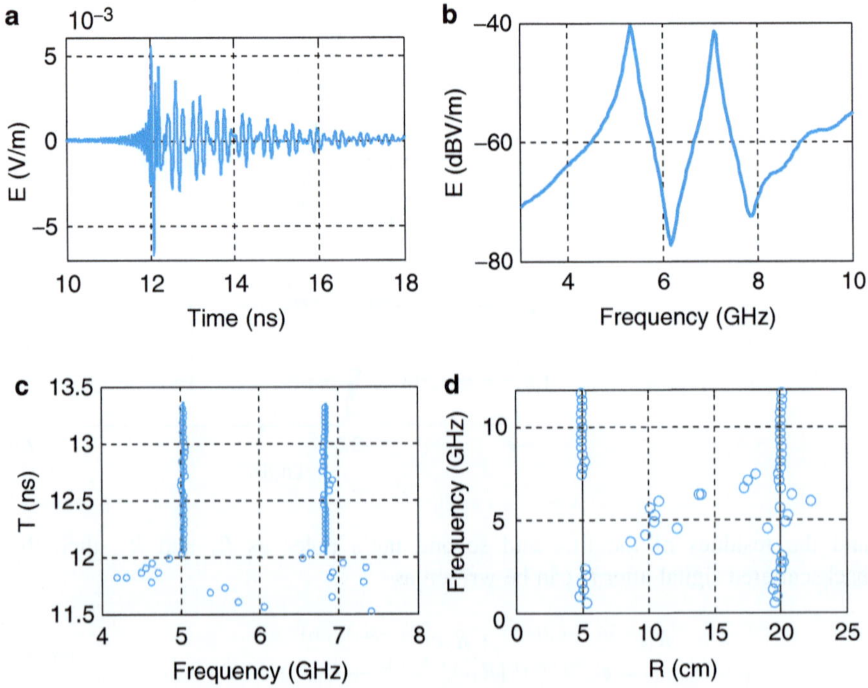

Fig. 5.8 (**a**) Time-domain signal (**b**) frequency-domain response (**c**) time-frequency diagram after STMPM (**d**) space-frequency diagram after SFMPM [3] with permission from IEEE

frequencies (ID = 00). In addition, the RCS of the second tag is much smaller than that of the first. The time-domain and frequency-domain responses are depicted in Fig. 5.8a, b. In this case, the early-time response of the second tag is obscured in the late-time response of the first tag. By applying STMPM and NFMPM to the time-domain and frequency-domain responses, the time-frequency and space-frequency diagrams of the signal are shown in Fig. 5.8c, d. Similar to previous examples, if we know the turn-on times and poles, the ID of each tag can be found by reconstructing the backscattered signal of each tag. Here, the presence of the second tag cannot be detected without the space-frequency diagram.

As another example, two 3-bit tags are assumed 20 cm apart. The configuration and dimensions of the tags are shown in Fig. 5.9. The tags represent ID_1:111 and ID_2:101. The presence and absence of each assigned resonant frequency represents bits 1 and 0, respectively. The assigned resonances are at $f_1 = 6.2$ GHz, $f_2 = 8.7$ GHz, and $f_3 = 10.9$ GHz. The simulation is performed in CST Microwave Studio. The pole diagram of the 3-bit tag is depicted in Fig. 5.10. The illuminating plane wave first hits tag #1 and then tag #2. The time, frequency, time-frequency and time-residue responses of the backscattered field are shown in Fig. 5.11. Obviously, the identification process cannot be done perfectly with backscattered frequency-domain response.

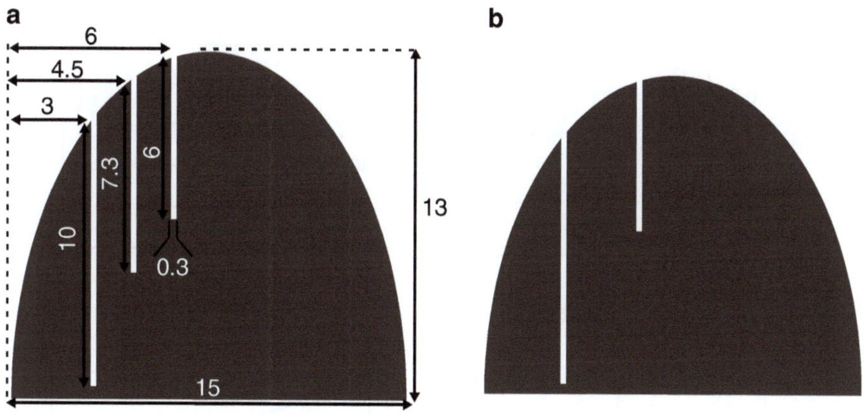

Fig. 5.9 Schematic view of the tags. (**a**) ID_1:111, (**b**) ID_2: 101 [3] with permission from IEEE

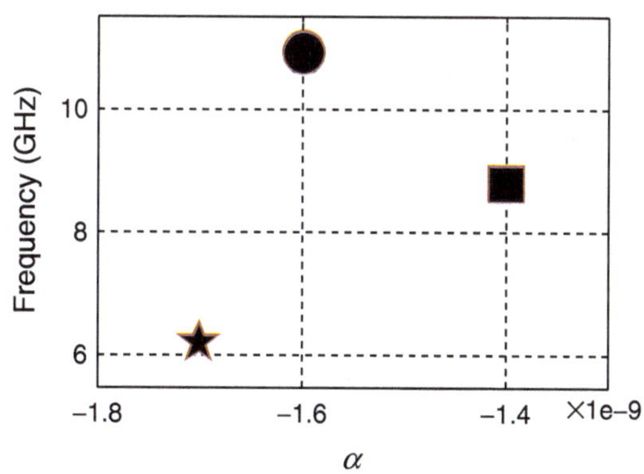

Fig. 5.10 Pole diagram of the 3-bit tag [3] with permission from IEEE

From the time-frequency response, it can be inferred that the first tag represents three resonant frequencies. At $t = 1.8$ ns, the illuminating wave hits the second tag with $ID_2 = 101$. According to Fig. 5.11c, three resonant frequencies exist in the backscattered signal after $t = 1.8$ ns. In the simulation results, the window length of $T_W = 0.5$ ns is used. Another important parameter introduced in matrix pencil method (MPM) is p which is the number of significant decimal digits in the sampled data. It acts as a filtering parameter determining the accuracy of the extracted poles. For noisy data, we usually use $p \leq 3$. Here, $p = 4$ is used for more accuracy. For values less than 4, just two poles of the second tag after $t = 1.8$ ns are represented in the time-frequency diagram. To identify the ID of the second tag, we use the time-residue diagram shown in Fig. 5.11d. As can be predicted by the mathematical formulation in (5.14) and (5.16), the residues of the first and third bits have jumps at the second turn-on time whereas the residues of the second bit are located in a straight line without any jump at the second turn-on time. It confirms that the second resonant frequency after $t = 1.8$ ns is related to the first tag's response.

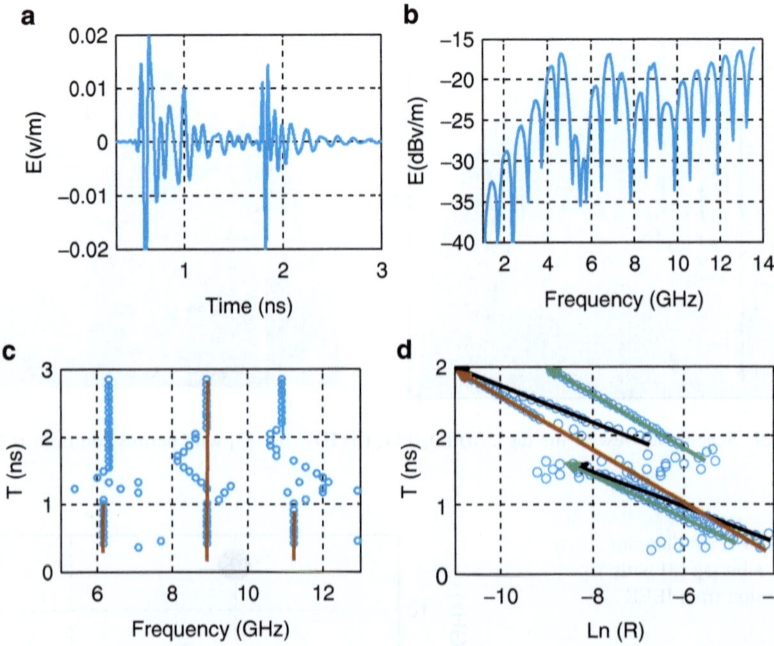

Fig. 5.11 (**a**) Time-domain backscattered signal from two tags spaced by $R = 20$ cm (**b**) frequency-domain response (**c**) time-frequency representation of the signal by applying STMPM with $T = 0.5$ ns. (**d**) Time-residue diagram of the signal [3] with permission from IEEE

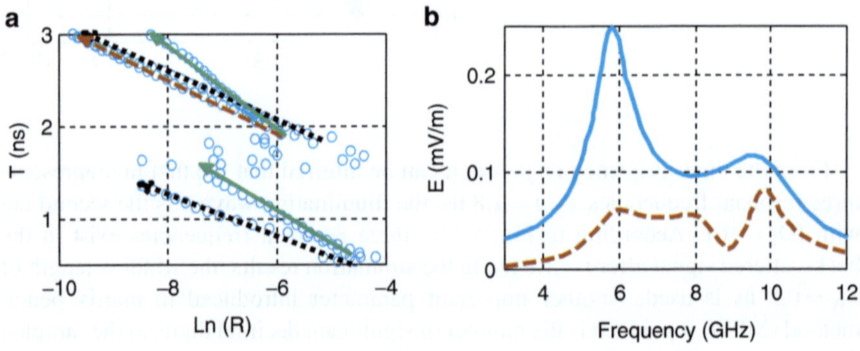

Fig. 5.12 (**a**) Time-residue diagram of the backscattered signal (**b**) separated responses of the tags in frequency-domain [3] with permission from IEEE

Therefore the ID of the second tag is $\mathrm{ID}_2 = 101$. The slopes of the poles in the time-residue diagram are related to the damping factors of the poles in Fig. 5.10. As an example, the second pole, which has the lowest damping factor, has the steepest slope in Fig. 5.11d.

As a final scenario, the incident wave first illuminates tag #2 with $\mathrm{ID}_2 = 101$, then second tag with $\mathrm{ID}_1 = 111$. The time-residue diagram of the backscattered signal is shown in Fig. 5.12a. In this case, all the residues have a jump at the second

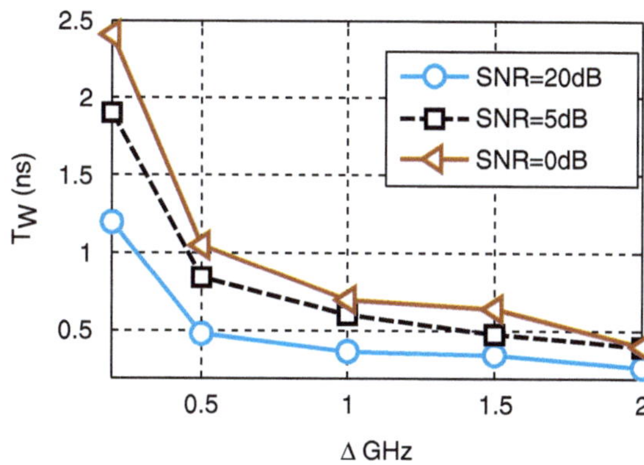

Fig. 5.13 Minimum required window length as a function of Δ for different SNRs [3] with permission from IEEE

turn-on time. By obtaining the residues and poles of the signal, the contribution of each tag to the backscattered signal is shown in Fig. 5.12b.

5.2.1 Space, Time and Frequency Resolutions

Similar to other time-frequency analysis methods such as wavelet transforms, short-time Fourier transform and so on, time and frequency resolutions are key parameters in the proposed method. The resolution in time and frequency domains is strongly dependent upon the length of the sliding window. The frequency resolution is related to the minimum distance in frequency between two adjacent resonant frequencies which can be identified. We discussed about the effect of window length and filtering parameter on the resolution in Chap. 4. Other important parameter affecting the frequency resolution is noise. In order to study the effect of the window length on the frequency resolution, the time-domain signal in (5.17) is considered as the backscattered signal from a 2 bit tag.

$$s(t) = Ae^{-\alpha t}\big(\sin(2\pi f_1 t) + \sin\big(2\pi(f_1 + \Delta)t\big) + n(t) \tag{5.17}$$

For more simplicity, the poles are assumed to have the same residue and damping factor and $n(t)$ is the additive Gaussian white noise. Assuming $f_1 = 5$ GHz and $\alpha = 1\text{e}8$ (Np/s), the minimum required frequency length of window for distinguishing the poles is shown in Fig. 5.13 as a function of Δ for different SNRs. As it shows, when the resonances of the tag are located closer to each other, we need to increase the window's length to distinguish the resonances. Also, for lower SNR cases, a larger value of T_w is required to distinguish the poles. The filtering parameter in MPM (p) is directly related to SNR. For lower SNRs, a lower value of p should be employed in the algorithm so as not to allow the poles due to noise to come into the picture. Here, the value of p is chosen as 5, 4, and 3 for SNR = 20, 5, and 0 dB, respectively. By increasing the length of time-window,

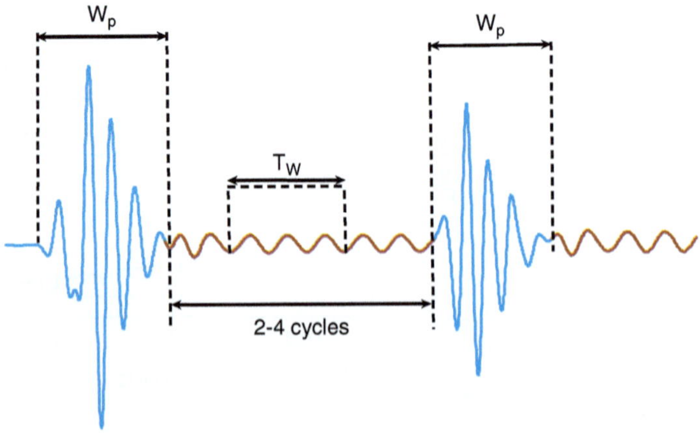

Fig. 5.14 Backscattered signal from two single-bit tags [3] with permission from IEEE

frequency resolution is increased which deteriorates the time resolution. For example, the turn-on time in Fig. 5.11c is located in between 1.5 and 2 ns which causes a significant error in calculating locations of the tags. On the other hand, the range resolution depends on the pulse width of the incident wave. Figure 5.14 shows the backscattered signal from two single-bit tags. The early-time response of the tags is a replica of the incident wave whose width can be approximated by [22]

$$W_{\mathrm{p}} = \frac{1}{B} = \frac{1}{(10.6 - 3.1)GHz} = 0.13ns \tag{5.18}$$

where B is the operational bandwidth of the incident wave (3.1–10.6 GHz). Hence, the range resolution can be calculated as [22]

$$\Delta R = \frac{cW_p}{2} = 2cm \tag{5.19}$$

In practical applications, we usually need a few cycles of the sinusoidal signals in the window to extract the poles. Considering the worst case as $f_{\min} = 3.1$ GHz, the minimum range resolution at which the tags and their IDs can be distinguished using the arrival times of the early-time responses is between 7 and 15 cm.

5.2.2 Measurement Results

Three 3-bit tags with IDs of $\mathrm{ID}_1 = 101$, $\mathrm{ID}_2 = 111$, and $\mathrm{ID}_3 = 011$ are designed based on the proposed technique in Chap. 3. The second and first bits of tags #1 and #3 respectively are nulled by soldering a stub in the middle of the related slots. This manual soldering causes a small shift in the resonant frequencies which is

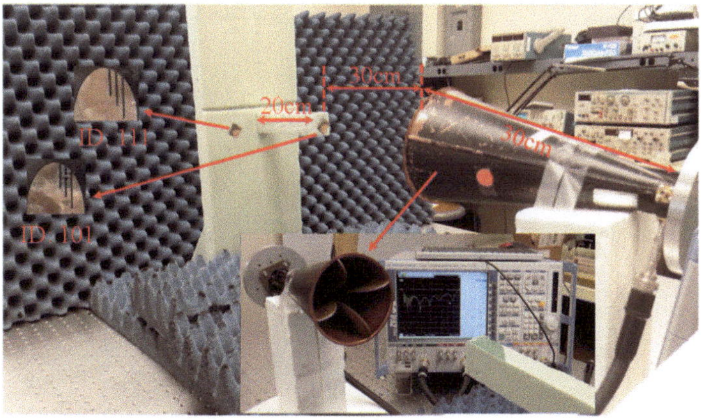

Fig. 5.15 Set-up for the measurement of backscattered signal from two tags [3] with permission from IEEE

Fig. 5.16 Time-domain backscattered signal from the tags [3] with permission from IEEE

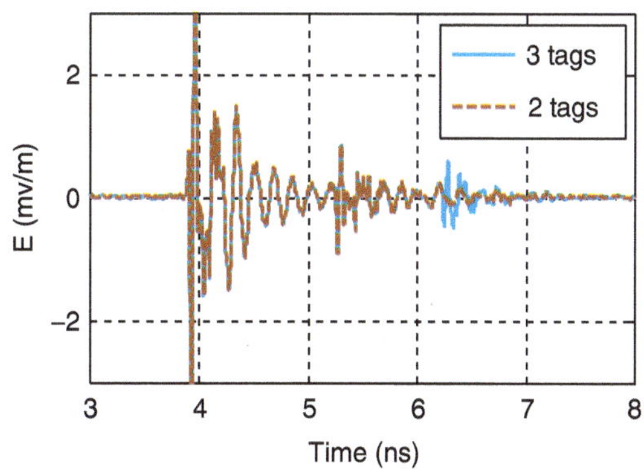

negligible in our analysis. Two measurements are performed: First, according to Fig. 5.15, two tags (ID_1 and ID_2) are located 20 cm away from each other in front of a UWB quad-ridge horn antenna. Then the third tag is located 15 cm from the second tag. The antenna is connected to the network analyzer and the S_{11} is measured while the frequency is swept from 10 MHz to 25 GHz. This wide frequency range is chosen for better resolution in the time domain. For practical applications, the standard frequency band (3.1–10.6 GHz) can be used for measurement. The data measured by the network analyzer cannot be used directly to extract the poles because it includes undesired components such as the contribution of TEM horn and the scattering from background objects in addition to the tag response. As a result, another measurement is performed without the presence of the tags to subtract the effect of the background objects from the first measured data. The input power is 18 dBm. The time-domain backscattered signal for two cases is shown in Fig. 5.16. By applying STMPM to the time-domain response, the time-frequency diagram of the signal is obtained, shown in Fig. 5.17. The optimum values of p and T_W are used at each snapshot of time for extracting the poles.

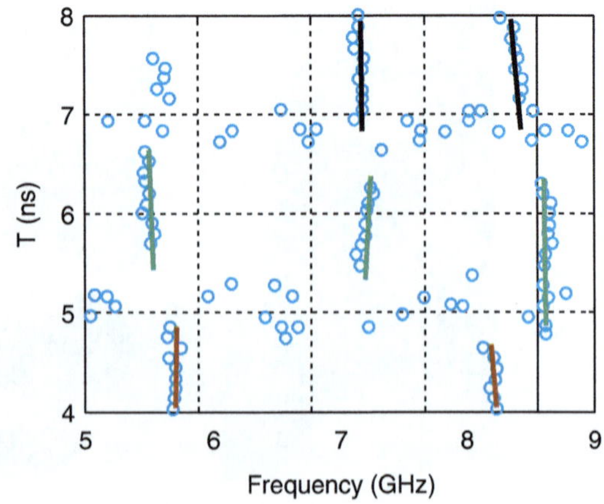

Fig. 5.17 Time-frequency representation of the backscattered signal [3] with permission from IEEE

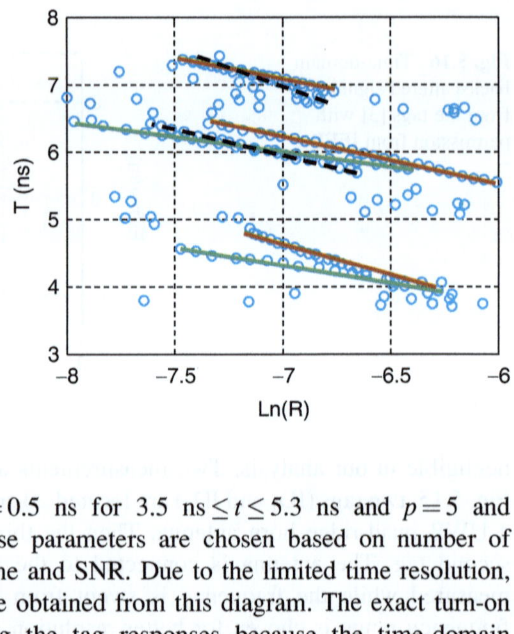

Fig. 5.18 Time-residue representation of the backscattered signal [3] with permission from IEEE

Here, we used $p = 4$ and $Tw = 0.5$ ns for 3.5 ns $\leq t \leq$ 5.3 ns and $p = 5$ and $Tw = 0.6$ ns for 5.3 ns $\leq t$. These parameters are chosen based on number of resonances in the snapshot of time and SNR. Due to the limited time resolution, the exact turn-on times cannot be obtained from this diagram. The exact turn-on time is crucial in reconstructing the tag responses because the time-domain reconstructed signal is very sensitive to phase (or turn-on times). As can be seen in Fig. 5.18, the second set of poles excited at $t = 5.3$ ns have different residues than the poles excited at $t = 3.9$ ns. Thus, the first and second tags represent two and three resonant frequencies, respectively. At $t = 6.2$ ns, the poles of the third tag are excited. The third tag is rotated 45° with respect to incident electric field. The resonant frequencies and ID of each tag can be identified from the time-frequency and time-residue diagrams. In Fig. 5.19, the real and imaginary parts of the backscattered signal versus frequency are shown. Following the algorithm presented in the flowchart of Fig. 5.3, by suppressing the late-time poles of the tags, the

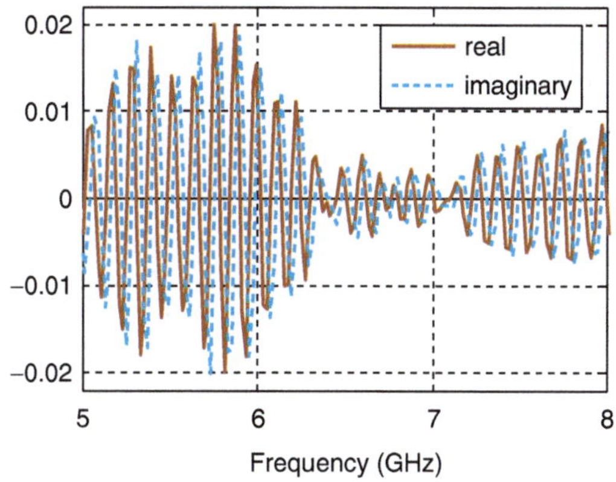

Fig. 5.19 Real and imaginary parts of the measured backscattered signal [3] with permission from IEEE

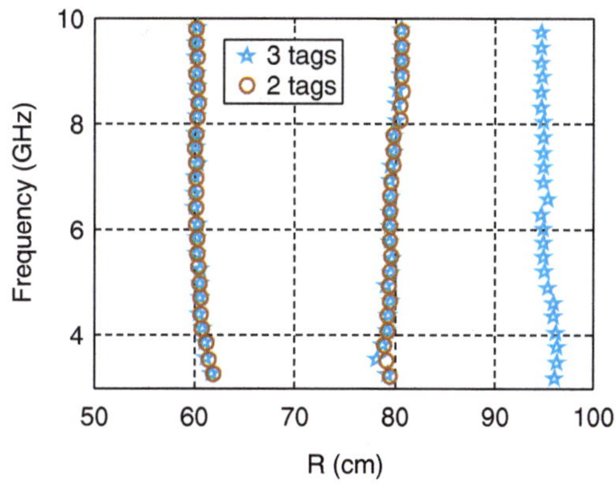

Fig. 5.20 Space-frequency diagram of the measured response [3] with permission from IEEE

space-frequency diagram of the response is shown in Fig. 5.20 for two cases. Here, the accurate turn-on times (or equivalently locations of the tags) are depicted in a space-frequency diagram.

5.3 Localization of Chipless RFID Tag

In tracking applications, it is desired to know the precise location of the tag in the reader area. For example, by knowing the tag location, the reader antenna can direct the antenna beam to the object, suppressing the interference signals from background objects [23]. In addition, by enabling this capability in conventional chipless RFID systems, it can be used in a wider range of crucial applications such as health-care monitoring in hospitals. For example, in microwave hyperthermia of breast cancer [24], the accurate localization of the tumor is necessary. Additionally, the localization technique is useful in the positioning of chipless RFID sensors placed in different locations of the medium in order to sense the

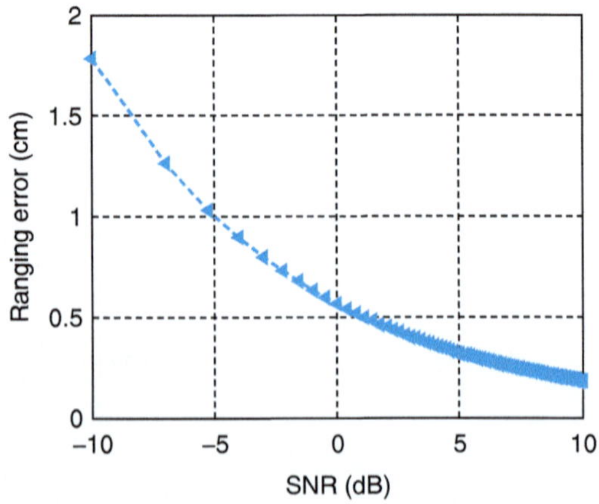

Fig. 5.21 Ranging error versus SNR [31] with permission from IEEE

density of a particular gas [25] or the humidity of the medium [26]. In such cases, because of the inhomogeneity of the material under consideration, multiple tags are used in different places. Therefore, the accurate localization and identification of the tags is an essential part of the measurement set-up.

Ranging techniques can be categorized as time-based ranging and received signal strength-based ranging [27]. The former is based on the time of arrival (TOA) of the signal while later is based on the principle that the greater the distance between two nodes, the weaker their relative received signals are. In practical applications, taking advantage of ultra-wideband technology in the detection process, the first method shows better accuracy and precision than the second [27]. However, there are some factors which affect the performance of the ranging process in the reader. In ideal propagation conditions, without considering multipath and interference phenomena in our discussion, the ranging accuracy for SNR larger than 15 dB is limited to the Cramer-Rao (CR) bound as [28, 29]

$$E\{R\} \geq \frac{c^2}{8\pi^2\beta^2 SNR} \tag{5.20}$$

Where $E\{.\}$ is the mean square error (MSE), R accounts for estimated range c is speed of light in free space and β represents the effective bandwidth of signal [28]. For lower SNRs, the estimation error is limited to a stricter bound, called the Ziv-Zakai bound [27]. The formulation is more difficult for multipath effects. As the simplest case, we consider the CR bound in (5.20). According to (5.20), the ranging error is strongly dependent on the SNR and the pulse shape. Figure 5.21 shows the ranging error as a function of SNR for CR bound in the presence of AWGN noise. The frequency bandwidth is 3.1–10.6 GHz. For SNRs lower than 10 dB, the Ziv-Zakai bound is more accurate which shows worse ranging error than the CR bound. It is seen that by decreasing SNR, the ranging error is increased. This may happen when the tag is located at larger distances from the reader antenna. In addition, the detection technique plays an important role in ranging calculations.

In most applications, classical matched filter (MF) TOA estimator is used to find the time when the signal has its maximum peak [27]. This strategy might not be the best method for localizing chipless RFID tags. The backscattered response from the tag includes early-time and late-time responses. In the cases where the early-time response is much stronger than the late-time response, the aforementioned technique works well. However in multi-bit tags, the late-time response of the tag is composed of high-Q sinusoidals corresponding to the embedded poles on the tag. At time instances when some sinusoidals are in-phase, their effect might be constructive enough to strengthen the late-time response at those time instances. Additionally, if two or more tags are located in the reader area, the early-time response of the second tag might be hidden in the late-time response of the first illuminated tag. Also, for bi-static cases, there is no guarantee that the early-time response is stronger than the late-time response.

Although many efforts have been made in the design and implementation of chipless tags, there is a demanding request for improved detection techniques in the reader, especially for localization applications. In [30], a space-time-frequency method has been suggested for the localization of the tags. Later on, the proposed method in [23] was confirmed by experimental results in [30]. The localization is based on the employment of three antennas spaced at different points in the reader area. For larger spaces, the area can be divided into some unit cells covered by some antenna arrangement. Although the authors in [23] did not consider the circumstances where the late-time response is stronger than the early-time part and also when multiple multi-bit tags are present in the reader area, the method performs well within 2.1 cm and 3.5° error in distance and angle for a single tag.

Here, a new technique is introduced for accurate localization of chipless RFID tags in the reader area. Similar to [30], three antennas are used in the unit cell. Assuming the reader area as the scattering area, the tags can be regarded as the scattering centers of the medium. Based on Altes' model, the early-time response from the reader zone can be expanded versus the localized impulse responses of the scattering centers. By applying NFMPM, which is the dual of the short-time matrix pencil method (STMPM) [31], to the frequency-domain response of the tag at each antenna port, its location can be easily found by some mathematical manipulations.

The major advantages of the technique proposed herein are as follows:

1. An easy-to-implement approach is proposed to qualitatively improve localization accuracy in chipless RFID systems.
2. This technique is applicable for localization of multiple multi-bit tags in the reader zone.
3. By obtaining the accurate value of turn-on time of the tag, its ID can be easily found by applying STMPM to the time-domain response.
4. By taking advantage of the three-antenna implementation in the unit cell, the effects of noise and polarization issues in the identification can be reduced considerably.
5. The interferences from adjacent unit cells can be strongly eliminated from the backscattered signals.

Figure 5.22 illustrates the system configuration arranged for the localization of chipless RFID tags. The area under consideration is subdivided into unit cells. Each unit cell is covered by three ultra wideband antennas spaced by 120° with respect to the center of the unit cell. The frequency of operation is assumed to be 3.1–10.6 GHz compatible with FCC requirements. By employing three antennas at each unit cell and obtaining the turn-on time of the tag (or equivalently the distances from the tag to the antennas), the position of the tag can be calculated easily. In addition, this arrangement has some other benefits which improves the detection and identification capability of the reader. By increasing the distance between tag and antenna, the amplitude of both early-time and late-time responses decrease which leads to a decrease in SNR. This results in two different drawbacks in the localization and identification of the tags. According to (5.20), by degrading SNR of the received signal, the ranging accuracy deteriorates. Also, by increasing the distance and adding extra noise to the signal, the extraction of the embedded poles of the tag becomes extremely challenging. With a three-antenna arrangement and some power considerations, we can ensure that the SNR of the backscattered signal from the tag is above a threshold for at least one of the antenna ports. Hence, we can obtain the signature of the tag by analyzing the strongest backscattered received signal from the tags. Furthermore, by receiving the backscattered signal from three different directions, the direction and polarization dependency of the tag can be mitigated. In conclusion, a three-antenna configuration in a unit cell facilitates both localization and identification processes in the reader part.

According to Fig. 5.22, three antennas in the unit cell are spaced at 120° on a circle of radius R. Assuming multiple multi-bit chipless RFID tags present in the reader area, the backscattered responses at the antenna ports contain the reflections from the tags, antennas and interferences from the adjacent unit cells. Considering $s_{\mathrm{i}}(t)$ as the input signal at the ith antenna port, the backscattered signal can be written as the combination of early-time and late-time responses of the scattering objects as

$$r_{\mathrm{i}}(t) = e_{\mathrm{i}}(t) + l_{\mathrm{i}}(t) = \sum_{n=1}^{N_{\mathrm{t}}+N_{\mathrm{o}}} \left[e_{\mathrm{n,i}}(t) + \mathrm{Re}\sum_{m=1}^{M_{\mathrm{n}}} R_{\mathrm{n,i}}^{\mathrm{m}} e^{-s_{\mathrm{n}}^{\mathrm{m}} t} \right] U(t - t_{\mathrm{n,i}}) \tag{5.21}$$

where $e_{\mathrm{n,i}}$ is the early-time response of the nth object at the ith antenna port. The second part in the bracket contains the late-time responses from the objects which based on singularity expansion method (SEM) is summed over all natural resonances ($s_{\mathrm{n}}^{\mathrm{m}}$) with weighting residues ($R_{\mathrm{n}}^{\mathrm{m}}$). $U(.)$ is the Heaviside function and $t_{\mathrm{n,i}}$ is the turn-on time of the *nth* scatterer at the ith antenna port. N_{t} is assumed to be the number of the tags in the cell, N_{o} is the number of signals other than the tag's reflections coming from background objects and interferences, and M_{n} is the number of bits embedded on the nth tag. Based on Altes' model, the early-time response from each scatterer can be expanded versus the impulse response of the localized scattering centers as

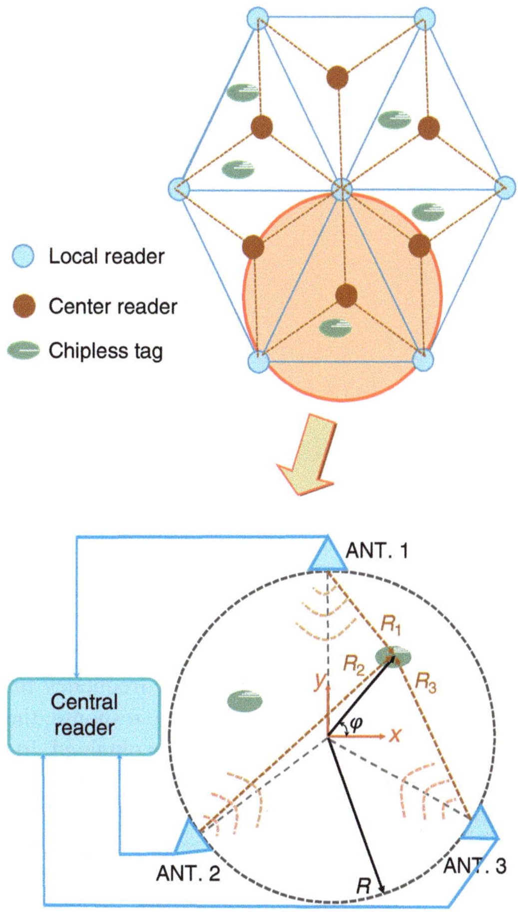

Fig. 5.22 System configuration for localizing chipless RFID tags in the reader area [31] with permission from IEEE

$$e_{\mathrm{n,i}}(t) = s_{\mathrm{i}}(t) * \sum_{p=-\infty}^{+\infty} a_{\mathrm{n,p,i}} \delta^{(p)}(t - t_{\mathrm{n,i}}) \tag{5.22}$$

where $t_{\mathrm{n,i}}$ is the delay-time from the ith antenna to the nth scatterer. The impulse response of the nth scatterer in (5.22) is summed over the integrals and derivatives of the Dirac-delta function.

Here, the negative and positive values of p refer to the pth integral and derivative of the delta function, respectively. For simple scatterers as in our case, one term of the summation might be enough. By inserting (5.22) in (5.21), the received signal in the Laplace domain can be written as

$$R_{\mathrm{i}}(s) = \sum_{n=1}^{N} \left[\sum_{n=-\infty}^{+\infty} A_{\mathrm{n,p,i}}(s) e^{-st_{\mathrm{n,i}}} + \sum_{m=1}^{M_n} \frac{R_{\mathrm{n,i}}^{\mathrm{m}}}{s - s_{\mathrm{n}}^{\mathrm{m}}} \right] \tag{5.23}$$

in which

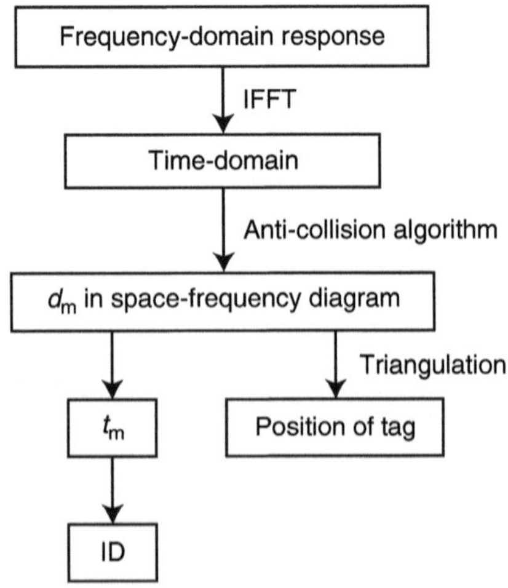

Fig. 5.23 Flowchart of proposed localization algorithm [31] with permission from IEEE

$$A_{n,p,i}(s) = a_{n,p,i} s^p S(s)$$

By employing the method proposed in Sect. 5.2, suppressing the late-time response and applying the narrow-frequency matrix pencil method (NFMPM) to the frequency-domain of early-time response, the accurate values of roundtrip time of the scattering centers, $t_{n,i}$, can be obtained. In this approach, a sliding-frequency window is moved along the frequency axis and the matrix pencil method is applied to each window. By converting the extracted complex times to distance as

$$d_m = \frac{c t_m}{2}$$

and plotting d_m versus sliding frequency, the distance from the tags to the antenna can be monitored in the space-frequency diagram. Knowing the distance of a tag from the antennas, its location can be calculated with respect to the reference point at the center of the unit cell. After obtaining the turn-on times of the tags, the time-window can be adjusted in the late-time response of the tag in order to extract the corresponding poles of the tag. The flowchart seen in Fig. 5.23 summarizes the proposed localization technique.

As an example, the 3-bit tag under consideration is shown in Fig. 5.24. It consists of three quarter-wavelength slots resonating at $f_1 = 5.1$ GHz, $f_2 = 7.1$ GHz, and $f_3 = 8.1$ GHz. The simulated and measured RCS of the tag is depicted in Fig. 5.25 when the incident electric field is perpendicular to slot length. Based on the RCS of the tag and sensitivity of the receiver, the input power can be adjusted so as to maintain an SNR above a certain threshold. For larger tags, the reflected signal is stronger which results in lower ranging error based on (5.20). According to [3], the minimum SNR at which the tag's ID can be accurately identified is about 15 dB.

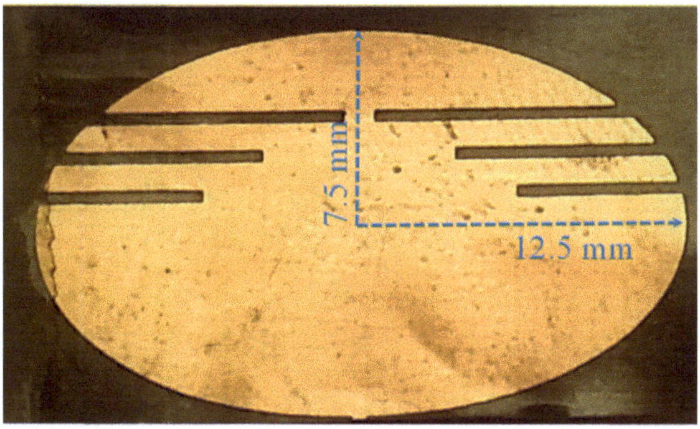

Fig. 5.24 Configuration of the 3-bit fabricated tag [31] with permission from IEEE

Fig. 5.25 Simulated and measured RCS of the tag when the incident electric field is perpendicular to slot length [31] with permission from IEEE

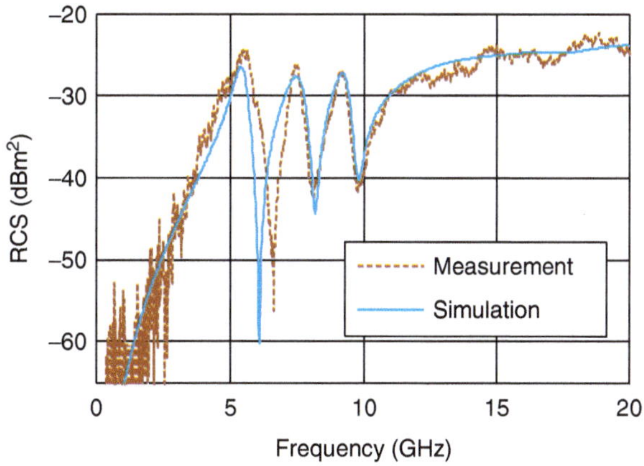

Assuming the tag is located in the far field of the antenna, the radar budget equation can be written in the frequency domain as

$$\frac{P_\mathrm{r}}{P_\mathrm{t}} = \frac{A^2 \delta(f) f^2}{4\pi c^2 R^4} \tag{5.24}$$

where P_r and P_t are the received and transmitted power at the antenna with the effective area of A. The quantity δ is the frequency-dependent RCS of the tag as shown in Fig. 5.25, c is the speed of light in free space and R is distance from tag to antenna. A TEM-horn antenna with a length of 30 cm and effective aperture surface area of 150 cm^2 is connected to a network analyzer to measure the backscattered signal from the tag. Assuming $R_{\mathrm{max}} = 100$ cm as the maximum detectable range, the ratio of received to transmitted power at the antenna is shown versus frequency in Fig. 5.26. The measured received noise by the antenna in the laboratory

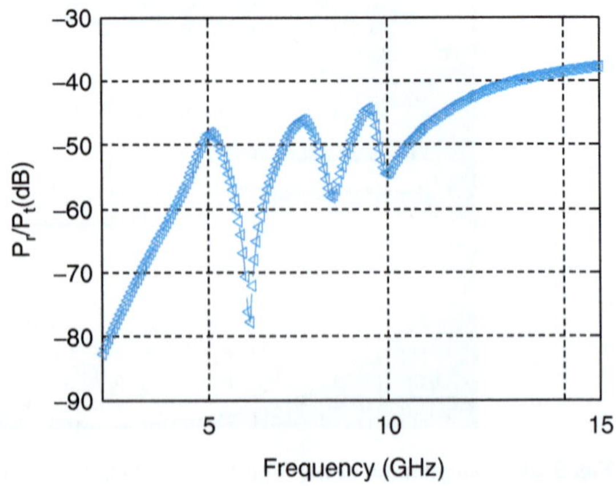

Fig. 5.26 Normalized received power at the antenna versus frequency [31] with permission from IEEE

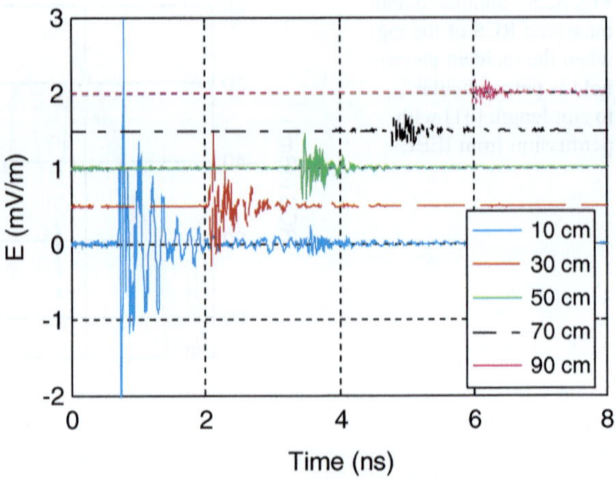

Fig. 5.27 Measured time-domain response from the tag for different distances [31] with permission from IEEE

environment is about −60 dBm. The peaks of the frequency-domain response in Fig. 5.26 are associated with the resonant frequencies of the tag. As it shows, at the first resonant frequency of $f = 5$ GHz, the value of P_r/P_t is −50 dB. This value is the combination of the early-time response and the corresponding residue of the late-time response of the tag at $f = 5$ GHz. For high-Q resonances, the late-time residue is effectively dominant. Assuming 10 dB for the early-time response and polarization mismatch, transmitted power should be adjusted to have $P_t > 10$ dBm in order to keep the SNR above 15 dB. By choosing $P_t = 20$ dBm, the backscattered response from the tag is calculated by subtracting the resultant S_{11} at the antenna port from the off-tag measured S_{11}. In Fig. 5.27, the measured backscattered electric field from the tag is depicted for different distances of the tag from the antenna. For better comparison, the electric fields are shifted up along E-axis. It is seen that by increasing the distance, the backscattered response becomes noisy. In order to extract the resonances of the tag from the backscattered signal, short-time matrix

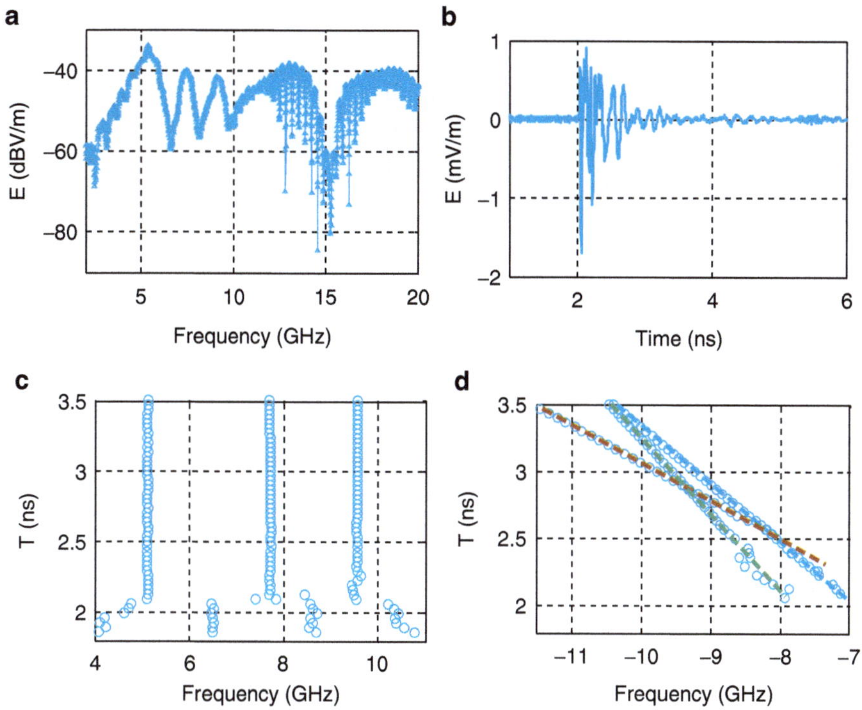

Fig. 5.28 (**a**) Frequency-domain, (**b**) time-domain, (**c**) time-frequency and (**d**) time-residue representation of measured backscattered signal from the tag [31] with permission from IEEE

pencil method (STMPM) is applied to the time-domain signal. As an example, the frequency-domain, time-domain, time-frequency, and time-residue responses of the backscattered signal from the tag located at 30 cm from the antenna are shown in Fig. 5.28. From the time-domain response, it is seen that $t = 2$ ns at which the signal has its maximum amplitude can be considered as the roundtrip time from the tag to the antenna. The time-domain response is composed of early-time and late-time responses. The early-time part emanates from the scattering centers of the tag and depends strongly upon the incident angle and polarization. The late-time response is the summation of damped sinusoidals corresponding to the slot poles. All the information embedded on the tag is included in the late-time response. The time-frequency diagram of the response shows the resonant frequencies (frequency-domain data) and also, the turn-on times (time-domain data) of the resonances in time. Because of limited time resolution in time-frequency analysis, the exact turn-on time cannot be extracted, especially when multiple tags are present in the reader area. It also leads to considerable error in ranging calculations. The real and imaginary parts of the backscattered signal are shown in Fig. 5.29 versus frequency. By applying NFMPM to the frequency response, the distance of the tag from the antenna is illustrated for different cases in a space-frequency diagram in Fig. 5.30.

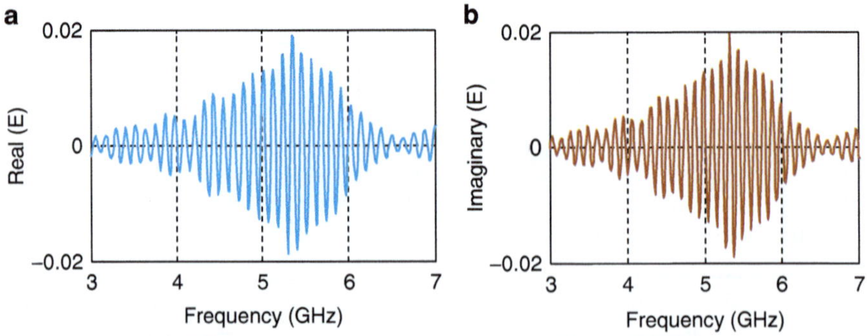

Fig. 5.29 (**a**) Real and (**b**) imaginary parts of the measured backscattered signal from the tag [31] with permission from IEEE

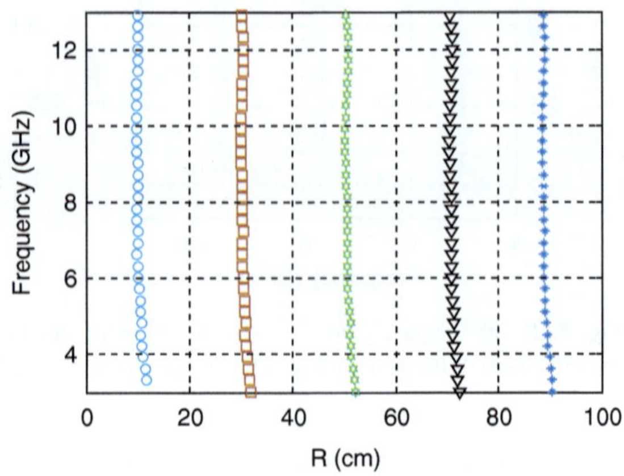

Fig. 5.30 Space-frequency diagram of measured backscattered response from the tag for different cases [31] with permission from IEEE

According to this diagram, the scattering centers, which in this case are the tags, reflect the incident pulse. The accuracy of the method is shown for different distances from tag to antenna.

The proposed method is applied to the cases where multiple tags are located in the main beam of the antenna. As an example, two 2-bit tags with high-Q resonators are spaced 25 cm apart. The scenario and dimensions of the tags are depicted in Fig. 5.31. An incident electric field illuminates the tags and the backscattered response is retrieved in the reader. The frequency and time-domain responses are depicted in Figs. 5.32 and 5.33 respectively for two rotation angles of the second tag. The tags carry two different IDs of 01 and 11, respectively. As the time-domain response shows, for $\varphi = 0$, the early-time response of the second tag is hidden in the late-time response of the tags. This phenomenon is more severe when more tags with higher density data are present in the reader zone. In Fig. 5.34, the space-frequency diagram of the backscattered response is shown for $\varphi = 0$, the worst-case scenario. Compared to the results shown in Fig. 5.30, the backscattered signals from the tags are normalized to the impulse response of the antenna. In the cases where the data is embedded in the spectral-domain response of the tags, the localization

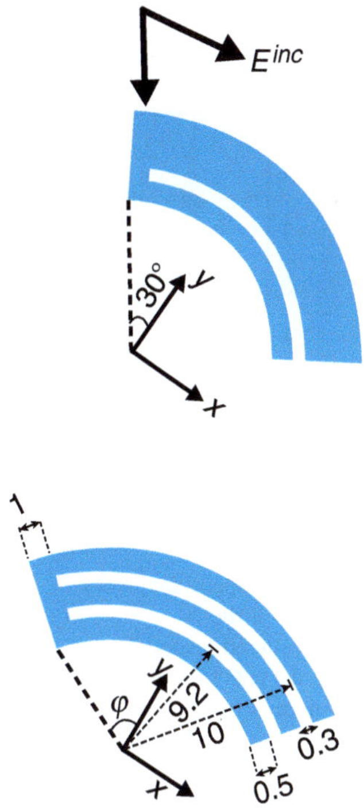

Fig. 5.31 Two 2-bit tags illuminated by an incident electric field. Units in mm [31] with permission from IEEE

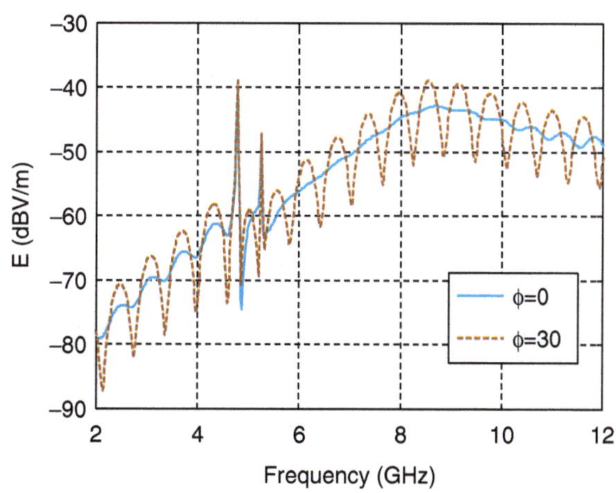

Fig. 5.32 Frequency-domain response of the backscattered electric field from two tags [31] with permission from IEEE

cannot be efficiently performed based on the time of arrival (TOA) or received signal strength (RSS) [23]. Meanwhile, by employing the proposed algorithm as shown in Fig. 5.23, not only can the location of the tags be obtained but their IDs can also be retrieved successfully.

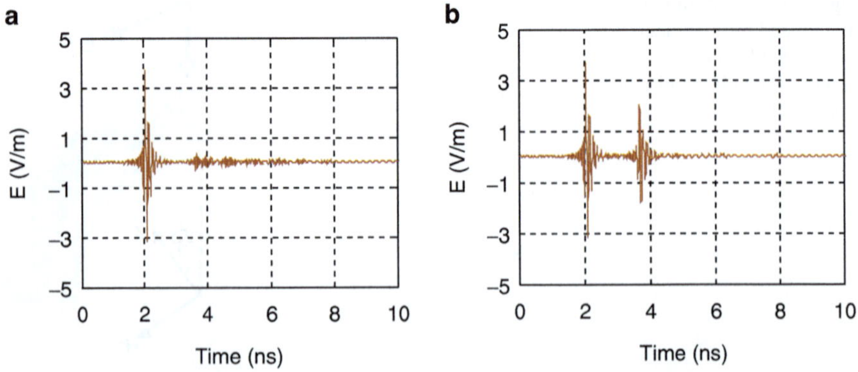

Fig. 5.33 Time-domain response from the tags for (a) φ = 0° and (b) φ = 30° [31] with permission from IEEE

Fig. 5.34 Space-frequency diagram of the backscattered response from the tags [31] with permission from IEEE

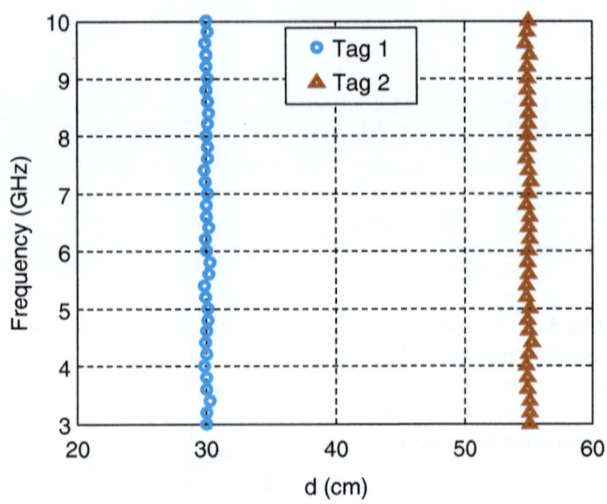

The aforementioned algorithm can be employed for finding the positions of the tags in the reader area. Here, we just consider one unit cell covered by three TEM horn antennas as Fig. 5.35 depicts. The antennas are interspersed along a circle of radius $R = 65$ cm. The simulation is performed in FEKO. The measurement set-up is shown in Fig. 5.36. Three TEM horns are connected to the network analyzer and two measurements in the presence and absence of the tag are performed. By taking the difference between two sets of measurements, the tag response is retrieved at the antenna ports. First, one tag is considered in the reader zone. By receiving the backscattered responses at the antenna ports and applying the proposed technique to the signals, the distances of the tag from three antennas are obtained. The position of the tag can be obtained via triangulation. In the cases where three circles do not intersect at a unit point because of the limited accuracy of the method, the closest point to the circles is considered as the tag position. For simplicity, we express the position of the tag in polar representation by (ρ, φ).

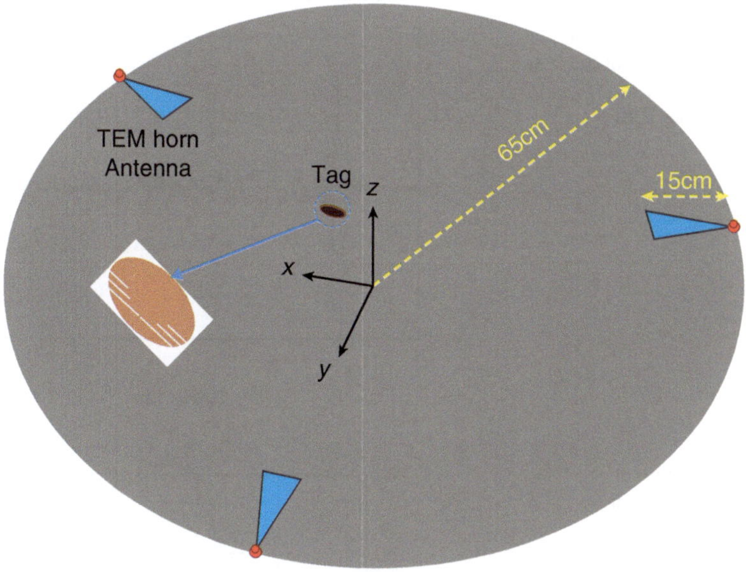

Fig. 5.35 Simulation set-up in FEKO [31] with permission from IEEE

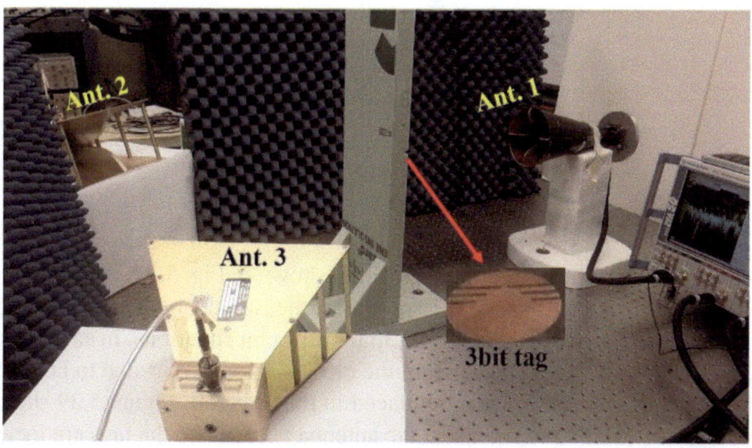

Fig. 5.36 Measurement set-up for localizing the chipless RFID tag [31] with permission from IEEE

Figure 5.37 shows the time-domain backscattered signals at the antenna ports when the tag is located at the center of the reader area perpendicular to the y-axis. It is seen that the strength-based positioning is not an accurate technique for localization of chipless RFID tags. Depending on the polarization and direction of the tag with respect to the antenna, the strength of the late-time and early-time responses may change. Hence, at some times the late-time response can be stronger than the early-time response which makes the localization more difficult.

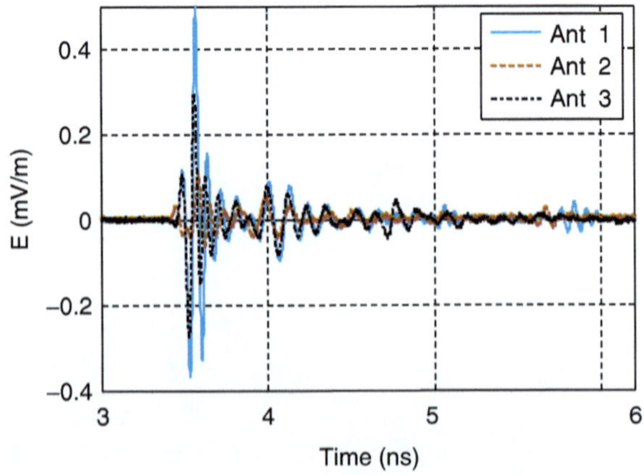

Fig. 5.37 Measured time-domain signals from the tag located at the center of the unit cell [31] with permission from IEEE

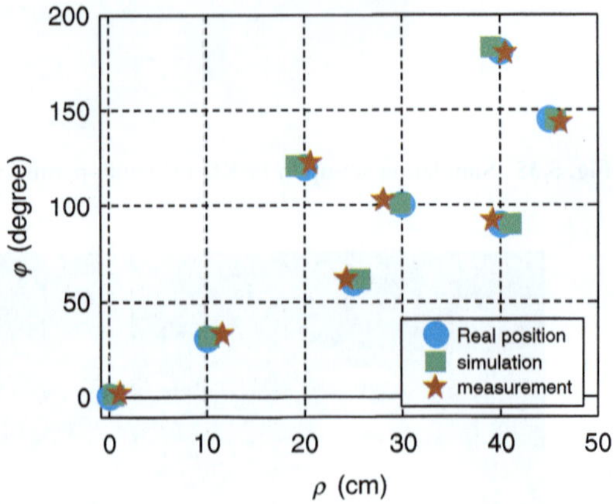

Fig. 5.38 Position of the tag extracted from the proposed technique compared to real position [31] with permission from IEEE

Instead, by employing the proposed algorithm shown in Fig. 5.23, the positions and IDs of the tags can be extracted. The simulated, measured, and real positions of the tag in the unit cell are depicted in Fig. 5.38 for different situations. In each situation, the ID of the tag can be extracted from the closest antenna port due to better SNR.

As another example, two tags are placed in the unit cell. Figure 5.39 shows the time-domain backscattered signals at the antenna ports when the tags are located at (20, 90°) and (20, 210°).

It is seen that for the first and second antennas where the second tag is further from the antenna, the turn-on time is not clearly visible. However, by applying the proposed technique, the positions of the tags can be extracted accurately. By applying short-time matrix pencil method (STMPM) to the time-domain response received at the first antenna port, the time-frequency response of the tags is shown in Fig. 5.40. The IDs of the tags are visibly indicated in Fig. 5.40 as ID_1:111 and ID_2:011. The accuracy of the proposed method is compared to the real positions of the tags in three different cases in Fig. 5.41. In each case, the locations of the tags are changed to compare the results.

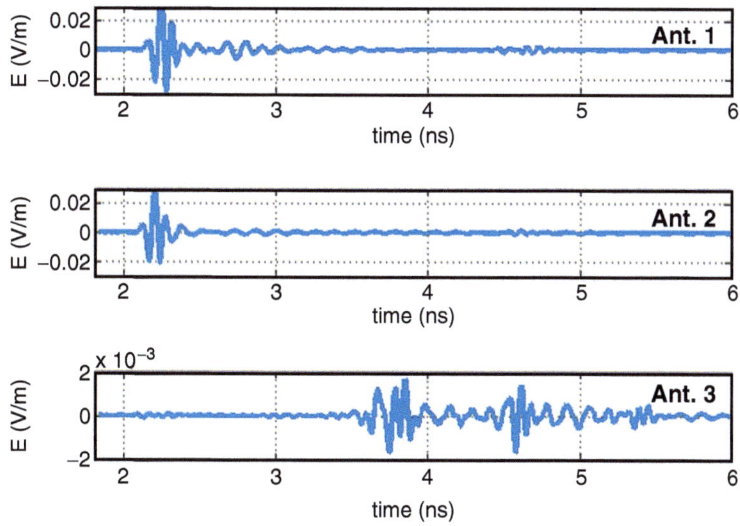

Fig. 5.39 Backscattered signals from two tags at the antenna ports [31] with permission from IEEE

Fig. 5.40 Time-frequency response of the received signal at the first antenna [31] with permission from IEEE

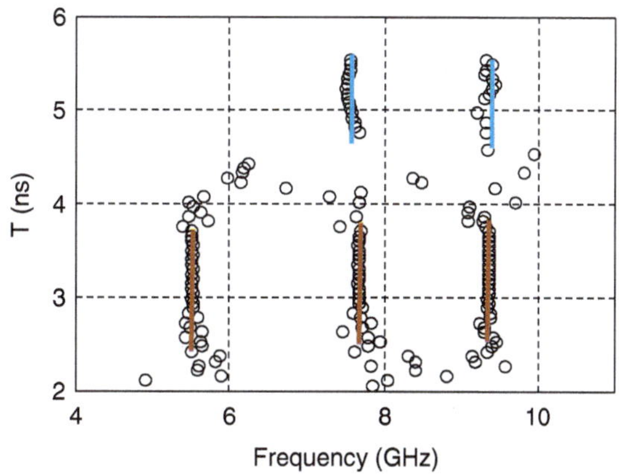

Fig. 5.41 Positions of the tags extracted from the proposed technique compared to real position [31] with permission from IEEE

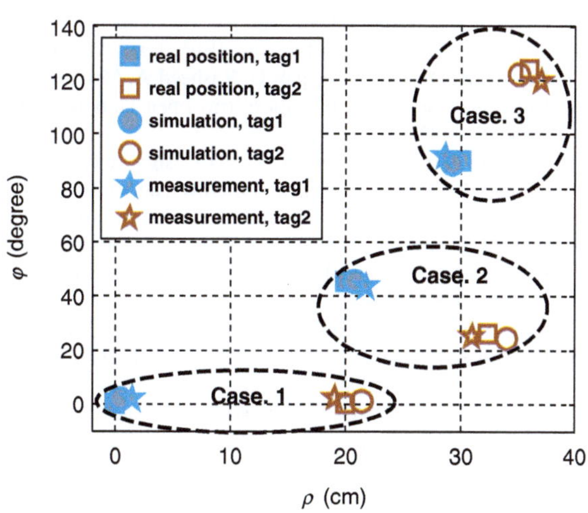

References

1. Azim R, Karmakar N (2011) A collision avoidance methodology for chipless tags. IEEE-proceeding Asia-Pacific microwave conference, pp 1514–1517
2. Azim R, Karmakar N (2012) Efficient collision detection method in chipless RFID system. International conference on electrical and computer engineering, pp 830–833
3. Rezaiesarlak R, Manteghi M (2013) A space-time-frequency anti-collision algorithm for identifying chipless RFID tags. IEEE Trans Antennas Propag 62:1425–1432
4. Rezaiesarlak R, Manteghi M (2013) Short-time matrix pencil method for chipless RFID detection applications. IEEE Trans Antennas Propag 61:2801–2806
5. Jin J-M (2010) Theory and computation of electromagnetic fields. Wiley, Hoboken, NJ
6. Baum CE (1971) On the singularity expansion method for the solution of electromagnetic interaction problems. Interaction Note 88, Air Force Weapons Lab., pp 1–111
7. Riley DJ, Davis WA, Besieris IM (1985) The singularity expansion method and multiple scattering. Radio Sci 20(1):20–24
8. Baum CE (1986) The singularity expansion method: background and developments. IEEE Antennas Propag Soc Newslett 28(4):14–23
9. Licul S, Davis WA (2005) Unified frequency and time-domain antenna modeling and characterization. IEEE Trans Antennas Propag 53:2882–2888
10. Altes RA (1976) Sonar for generalized target description and its similarity to animal echolocation systems. J Acoust Soc Am 59:97–106
11. Li Q, Rothwell EJ, Chen KM, Nyquist DP, Ross J, Bebermeyer R (1994) Determination of radar target scattering centers transfer function from measured data. Antenna and Propagation Society International Symposium, pp 2006–2009
12. Li Q, Ilavarasan P, Ross JE, Rothwell EJ, Kun-Mu C, Nyquist DP (1998) Radar target identification using a combined early-time/late-time E-pulse technique. IEEE Trans Antennas Propag 46:1272–1278
13. Heyman E, Felsen LB (1985) A wavefront interpretation of the singularity expansion method. IEEE Trans Antennas Propag 33:706–718
14. Hong SK, Wall WS, Andreadis TD, Davis WA (2012) Practical implementation of pole series convergence and the early-time in transient backscatter. Naval research laboratory, NRL/MR/5740-12-9411
15. Blischak AT, Manteghi M (2011) Embedded singularity chipless RFID tags. IEEE Trans Antennas Propag 59:3961–3968
16. Manteghi M, Rahmat-Samii Y (2007) Frequency notched UWB elliptical dipole tag with multi-bit data scattering properties. In Antennas and Propagation Society International Symposium, 2007 IEEE, San Diego, CA, pp 789–792
17. Preradovic S, Balbin I, Karmakar NC, Swiegers G (2009) Multiresonator-based chipless RFID system for low-cost item tracking. IEEE Trans Microw Theory Tech 57(5):1411–1419
18. Nijas CM, Dinesh R, Deepak U, Rasheed A, Mridula S, Vasudevan K et al (2012) Chipless RFID tag using multiple microstrip open stub resonators. IEEE Trans Antennas Propag 60:4429–4432
19. Vena A, Perret E, Tedjini S (2012) Design of compact and auto-compensated single-layer chipless RFID tag. IEEE Trans Microw Theory Tech 60:2913–2924
20. Costa F, Gnovesti S, Monorchio A (2013) A chipless RFID based on multiresonant high-impedance surfaces. IEEE Trans Microw Theory Tech 61(1):146–153
21. Rezaiesarlak R, Manteghi M (2013) A low profile multi-bit chipless RFID Tag. Radio Science meeting (USNC-URSI)
22. Skolnik M (2001) Introduction to radar systems. McGraw-Hill, New York
23. E-Azim R, Karmakar NC (2013) Chipless RFID tag localization. IEEE Trans Microw Theory Tech 61(8):2982–2994
24. Hong SK, Davis WA (2013) Use of tumor-specific resonances for more efficient microwave hyperthermia of breast cancer. Microw Opt Technol Lett 55(11):2659–2665

25. Vena A, Sydanheimo L, Tentzeris MM, Ukkonen L (2013) A novel inkjet printed carbon nanotube-based chipless RFID sensor for gas detection. Proceeding of the 43rd European microwave conference
26. Nair RS, Perret E, Tedjini S, Baron T (2013) A group-delay-based chipless RFID humidity tag sensor using silicon nanowires. IEEE Antennas Wirel Propag Lett 12:729–732
27. Dardari D, Conti A, Ferner U, Giorgetti A, Win MZ (2009) Ranging with ultrawide bandwidth signals in multipath environments. Proc IEEE 97(2):404–426
28. Dardari D, D'errico R, Roblin A, Sibille A, Win MZ (2010) Ultrawide bandwidth RFID: the next generation? Proc IEEE 98(9):1570–1582
29. Chung WC, Ha DS (2003) An accurate ultra wideband (UWB) ranging for precision asset localization. IEEE conference on Ultra wideband systems and technologies, pp 389–393
30. Manteghi M (2011) A space-time-frequency target identification technique for chipless RFID applications. In Antennas and Propagation Society International Symposium (APSURSI), 2011 IEEE, pp 1–4
31. Rezaiesarlak R, Manteghi M (2014) A space-frequency technique for chipless RFID tag localization. IEEE Trans Antennas Propag 62(11)

MIX
Papier aus verantwortungsvollen Quellen
Paper from responsible sources
FSC® C105338

If you have any concerns about our products,
you can contact us on
ProductSafety@springernature.com

In case Publisher is established outside the EU,
the EU authorized representative is:
Springer Nature Customer Service Center GmbH
Europaplatz 3, 69115 Heidelberg, Germany

Printed by Libri Plureos GmbH
in Hamburg, Germany